TURNING
TO
STONE

ALSO BY MARCIA BJORNERUD

Reading the Rocks: The Autobiography of the Earth

*Timefulness: How Thinking Like a
Geologist Can Help Save the World*

Geopedia: A Brief Compendium of Geologic Curiosities

TURNING
TO
STONE

*Discovering the Subtle
Wisdom of Rocks*

Marcia Bjornerud

Illustrations by Haley Hagerman

FLATIRON
BOOKS
NEW YORK

www.flatironbooks.com

Designed by Susan Walsh

Title page illustration by Viktoria Raikina / Shutterstock
Illustrations by Haley Hagerman

Library of Congress Cataloging-in-Publication Data

Names: Bjornerud, Marcia, author.
Title: Turning to stone : discovering the subtle wisdom of rocks /
    Marcia Bjornerud.
Description: First edition. | New York : Flatiron Books, 2024. |
    Includes bibliographical references and index.
Identifiers: LCCN 2023059084 | ISBN 9781250875891 (hardcover) |
    ISBN 9781250875907 (ebook)
Subjects: LCSH: Bjornerud, Marcia. | Women geologists—United
    States—Biography. | Geologists—United States—Biography. |
    Geology—Popular works. | Petrology—Popular works. | Geological
    time—Popular works.
Classification: LCC QE22.B54 A3 2024 | DDC 551.092 [B]—dc23/
    eng/20240316
LC record available at https://lccn.loc.gov/2023059084

Our books may be purchased in bulk for promotional, educational, or business use. Please contact your local bookseller or the Macmillan Corporate and Premium Sales Department at 1-800-221-7945, extension 5442, or by email at MacmillanSpecialMarkets@macmillan.com.

First Edition: 2024

10 9 8 7 6 5 4 3 2 1

For O, F, K, J, G, P, and N

CONTENTS

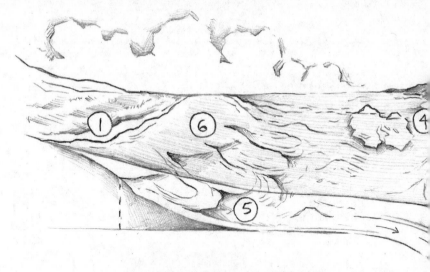

ITINERARY FOR A JOURNEY THROUGH EARLIER WORLDS

Original Geologic Settings
of Rocks in Each Chapter

1. **Sandstone:** Oceanside beach
   (Wisconsin, 500 million years ago)
2. **Basalt:** Continental rift
   (Lake Superior region, 1 billion years ago)
3. **Tuff:** Volcanic caldera
   (Long Valley, California, 800,000 years ago)
4. **Diamictite:** Glacial fjord
   (Svalbard, 700 million years ago)
5. **Turbidite:** Deep-ocean floor
   (Ellesmere Island, 400 million years ago)
6. **Dolomite:** Continental shelf
   (Ohio, 450 million years ago)

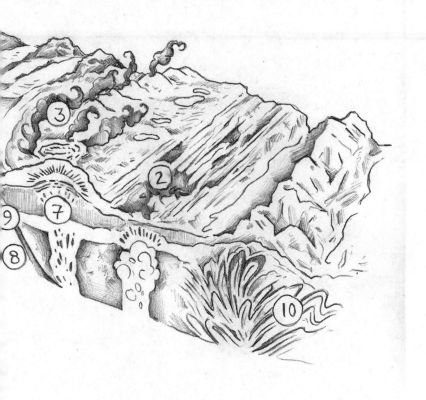

7. **Granite:** Continental volcanic arc
   (Wisconsin, 1.7 billion years ago)
8. **Eclogite:** Subduction zone
   (western Norway, 420 million years ago)
9. **Glass and flint:** Accretionary prism
   (South Island, New Zealand,
   100 million years ago;
   central Italy 10 million years ago)
10. **Quartzite:** Collisional mountain belt
    (Wisconsin, 1.5 billion years ago)

TURNING
TO
STONE

PROLOGUE

Ice

Viewed from the rails of a heaving ship, Svalbard conveys monumental stillness. When I first glimpsed the ice-draped islands, they seemed like a scene from the past, frozen in time, insensible to the din of civilization.

Initial explorations on foot confirmed this interpretation; no motion could be detected in the landscape. The wind was relentless, but on the tundra, no swaying trees or rippling grasslands made its power visible. We peered anxiously through binoculars at a patch of white in the middle distance, fearing it was a polar bear, only to realize it was a huge, cream-colored boulder that had been sitting in the same spot while the whole of human history elapsed.

On occasion, we'd hear strange noises—rumbling, crashing, roaring—and would scan the terrain to see what had produced them. But the land betrayed no hint of unrest. Days later, we'd find newly scoured channels on the mountainsides with levees of large rocks, testaments to avalanches that must have been the source of the thunderous sounds.

My companions and I were geologists in training, graduate students charged with deciphering the tectonic evolution of the northern end of the Appalachian-Caledonian mountain chain, the hemisphere-scale belt of crumpled rocks formed by continental collisions during the assembly of Pangaea.

This remote Norwegian archipelago, now alone in the Arctic Ocean, had once been attached to a sprawling supercontinent. While mapping the rocks as we found them in the present, we had to hold in our minds the geographies of the past.

Intellectually, we all understood that landscapes are impermanent, that the modern topography of Svalbard is a legacy of the Ice Age, that long before Pangaea existed, the contorted rocks we were studying had been flat layers of sediment on an ancient sea floor. But as flesh-and-blood mortals laboring up slopes toward the formidable peaks, we found it difficult to imagine that they had not been there forever. Too often, as we struggled to understand the mountains' interior architecture, they seemed not only unyielding but also hermetic, closed to interpretation. Glacial ice, equally mute, conspired with them to conceal their secrets.

Events that had shaped the landscape more recently were easier to read from the topography. The broad strand plain, which slopes gently from the base of the mountains to the sea, was clearly once a beach, beveled off by waves. The old shoreline is now covered by a layer of springy moss and peat, but as we walked across it on our daily commute into the mountains, we could feel the old wave-rounded cobbles rolling beneath our feet. The present elevation of this surface, a few tens of feet above sea level, reflects the rebound of the land since the late Pleistocene, when all of Svalbard was under a thick blanket of ice. As the Ice Age glaciers melted, sea level rose dramatically, but in the last ten thousand years—the duration of human history—the land has bounced back even more, lifting the former beach beyond reach of the raging surf. Perhaps Svalbard was not as still as it seemed.

One day, high in the mountains, staggering under packs

full of rock samples, we realized that a valley glacier might provide a shortcut back to base camp. The ice looked passable from where we stood—blue, not white, a good sign that there would be no crevasses hidden by last year's snows. There was a chance that we'd have to turn back if we encountered a deep meltwater canyon in the ice, or between the glacier and valley wall on the opposite side, but in our exhausted state, we decided to take the gamble. We buckled spiky crampons onto our boots, looped the straps of ice axes around our wrists, and started crunching across the glacier's rough surface.

At about the halfway point, we hopped over a small trench in the ice and were about to continue onward when we heard a gurgling noise that rose in pitch, like a drink being poured into a tall glass. Then came the sound of rushing water, and we turned to see that the channel we had just jumped across was now a small torrent. Thirty seconds later, it was empty again. We exchanged puzzled looks and then heard the gurgling once more from upstream, followed by another surge of water in the rivulet. We watched in fascination as this cycle repeated itself over and over, the water in the channel pulsing every minute or so like the icy lifeblood of the glacier.

Then we understood: Higher up on the ice, the little river ran through a series of pools, each a bit deeper than the main channel. When the topmost pool overflowed, another freshet would be sent downstream, and each pool would fill and empty in succession until the one just above us was overtopped and water spilled into the channel we had crossed.

We all laughed; it felt as if the glacier had told a joke, and that somehow, improbably, we humans had grasped the humor. Understanding the glacier's logic created an odd feeling of intimacy with it. Although the ice still obscured critical rock outcrops and posed dangers to careless geologists, it no longer seemed coldly incommunicative. We had a sense not only of its physical character but also of something like its personality. The glacier did grant us safe passage back to camp that day.

In the weeks that followed, I began paying more attention to the habits of glaciers and discovered other instances of their wit and creativity. In a narrow rock canyon, huge boulders were wedged tight high above the valley floor as if levitating, placed there by ice like whimsical pieces in a colossal sculpture garden. Crossing an inland ice field one day, I encountered a dozen narrow streams of cold air—not wind, but separate density-driven currents gliding silently downslope. Traversing them was like passing through a series of invisible doorways into and out of rooms with different thermostat settings. Such elaborate installations seemed to call for appreciative audiences.

Glaciers also modeled for us, in miniature, how strata become wrinkled in mountain building. Glacial ice, built up over millennia when summer warmth does not melt winter snow, is technically a kind of metamorphic rock—a crystalline substance altered by burial and pressure.

Fine layers of dust in the ice formed swirling patterns reminiscent of marble, and we could see that these convolutions had formed as the glaciers flowed languidly down their valleys. This made it easier to believe that the contorted strata visible on the mountainsides (thanks to the scouring ice) had formed in a similar manner: solid rock too could flow, given enough time.

During that first Arctic summer, I began to understand that the apparent stillness of the Svalbard landscape was completely illusory. Everything was in motion: slopes were slip-ping, ice was pulsing, the land was rebounding, the mountains were flowing, the geography of the world was changing.

It took some effort not to revert to my quasi-mystical childhood perception of the natural world, in which plants, rocks, and streams seemed to be whispering to one another. I grew up in a rural part of Wisconsin in the 1970s, an odd and awkward girl who found trees and stones to be good companions. Like most children, I simply assumed that objects were sentient, had preferences and personalities. As an aspiring scientist, however, I had trained myself to set any such notions aside in order to develop an objective, analytical understanding of nature. Surviving graduate school requires that one become an orthodox adherent to the rules of one's academic sect, and in the sciences, one of the deadliest sins is animism—describing inert entities like glaciers and moun-tains as if they were living beings. And yet, in Svalbard it was undeniable that the austere landscape *was* alive, its rocks and water, ice and air in constant conversation. The terrain was animate, sentient, and creative. It would just take me thirty years to say that out loud.

Turning to Stone is the story of my deepening relationship with our rocky planet, as a geologist whose career has coincided with an extraordinary period of scientific discovery. If the golden age of physics occurred in the first few decades of the twentieth century, with discoveries about the struc-ture of the atom and Einstein's insights on matter and energy, the glory days of geoscience began far more recently—and are in full flower today. When I first encountered the field in the late 1970s, conceptual frameworks for understanding

Earth at the planetary scale were just emerging: the theory of plate tectonics was barely a decade old, climate science was in its infancy, and the notion of global biogeochemical cycles was new. Geology was still a collection of non-intersecting subfields—mineralogy, petrology, sedimentology, paleontology, geomorphology—with limited views of the planet and research agendas driven mainly by the hunt for fossil fuels and mineral deposits. But in the early 1980s, new instrumentation, greater computational capacity, and a shift toward systems thinking started to change the discipline. In a beautiful irony, the grimy search for coal, oil, and ores had begun to reveal a stunning view of Earth across the eons.

And yet, few average Earth citizens have had access to the sweeping vision of the planet through time that modern geoscience now provides, and to those who have never been taught the language of the Earth, its story seems inaccessible and irrelevant. The seductive charms of technology and capitalism, in turn, have led to the delusion that we humans have invented the world and that the planet is just a mute, impassive backdrop to the "real" action. Such self-absorption is dangerous, for environmental, psychological, and even spiritual reasons.

In the popular imagination, physics is perhaps the science most often associated with deep philosophical, quasi-religious questions. The Higgs boson, for example, is informally known as the "God particle," and great physicists have the status of secular high priests. But for ordinary people, cosmological concepts are too abstract to understand in any visceral way; modern physics is largely nonphysical, at least in any normal human experience of the physical. In contrast, geology, with its focus on tangible records of the distant past, offers a bridge between human experiences of the world and the awe-inspiring but cold and formidable emptiness of space. Learning to read the storylines of Earth's history directly from rocks—understanding the plots and protagonists that shaped the places where we live—can help to provide a feeling of "embeddedness" in the cosmos, a sense of continuity and kinship with past and future. Perhaps the most distinctive characteristic of geologic thinking is the practice of roaming freely across many scales in space and time. In doing so, we can see ourselves in miniature, part of a long lineage of creatures on a creative planet that has renewed itself for more than four billion years while keeping an idiosyncratic diary of its activities over time in the form of rocks.

This book is an invitation into my geocentric worldview in which rocks are raconteurs, companions, mentors, oracles, and sources of existential reassurance. If my initial relationship with geology was purely intellectual, it has become over time philosophical and even transcendental, imbuing my life with meaning. My work as a field geologist has led to unlikely connections with people from across the world, friendships rooted in a shared sense of humility and wonder in the presence of nature. I also have a feeling of amity with rocks, after spending such a large fraction of my waking hours with them over many years, immersed in their narratives.

To me, rocks have distinct personalities. Each type has its own idiom. Some are extroverts, easily read and eager to share, while others are reticent and cryptic. Some are like celebrities, famous and beautiful but of little functional importance to the world, while many that seem plain and dull are essential to the long-term stability of the planet. Each chapter in this book features a rock type that was a protagonist at a certain time in my life, from an insular girlhood in northern Wisconsin, through an unlikely vocation as a geologist studying mountain-building processes in remote parts of the world. With each chapter, the stories of the rocks become more complex, and my relationship with them more intimate. Before I began to study geology, rocks were no more than a dimly perceived presence. But as I gradually learned to interpret their messages, they became embedded deep in my psyche. Now I have rocks in my head.

In sharing my small human story, I hope to convey the grander dramas embodied in Earth's pantheon of rocks—and to show how rocks, whether one ever thinks about them or not, ultimately define reality for everyone who lives on this rocky planet.

As I have experienced different life roles—daughter, sister, friend, wife, mother, professor, widow—and faced inevitable change and loss, the company of rocks has been a constant. Their *gravitas* is deeply comforting. Their stories are mythic in scope but undistorted by euphemism, delusion, or self-promotion: a bracing counterpoint to the deceptions, intentional and otherwise, that permeate everyday life. I believe that developing a collective sense of ourselves as Earthlings—native inhabitants of an old, durable planet—may bring reassurance in a time when so many human systems that once seemed robust are showing signs of fragility.

Although Earth started with the same raw materials as its siblings, this planet has recombined those ingredients into rocks found nowhere else in the solar system and invented processes that continuously recycle and regenerate them. In the face of such vigor and durability, the only rational response is to conform to the planet's ways of doing things.

While not everyone can or should become a geologist, geologic habits of mind—an instinct for Earth's rhythms, a feeling for our place in its story—are essential to the physical and psychic well-being of humankind. Around the time of the Industrial Revolution, Western society stopped seeing the Earth as imbued with wisdom—offering, in Shakespeare's words, "books in running brooks, sermons in stone"—and demoted it to dumb matter that we could outwit and exploit. This view of Earth as inert and unresponsive has led not only to environmental catastrophe but also to cultural anomie: We don't remember who we really are.

Contrary to their reputation, rocks are alert, responsive, communicative. They are the force field in which we live, the infrastructure of our existence, playing central roles in the "economy" of the world, as aquifers, architects of topography, managers of ocean chemistry, regulators of climate. They are also shape-shifters and fortune tellers. Once one becomes attuned to the language of rocks, it is obvious that Earth is vibrantly alive—and speaking to us all the time. Although we have sickened ourselves and vandalized our own home through ignorance and hubris, we may be restored by turning to stone and heeding its lessons in durability. Let me introduce you to some of the rocks that have helped me understand what it means to be an Earthling.

1

SANDSTONE

Even now, the impatient sound of an idling school bus makes my stomach knot. That rumble evokes a warm September morning in northwestern Wisconsin. It's 1970—the year of the My Lai massacre, the Kent State shootings, the first Earth Day—and I'm about to turn eight. I can't see the bus through the trees, but I hear the engine's *thrum-thrum-thrum*, like giant fingers drumming, and know it's already waiting at the end of the unpaved lane leading from our house deep in the woods. So I race as fast as I can, clutching my lunch bag and library book, bare knees pumping beneath my skirt, new shoes pinching my summer feet, but the soft sand slows me down.

I reach the county road just as the driver is about to turn the bus around and leave. Neighbor kids are already settled on board. Red-faced from running, I flush still more at the double indignities of almost missing the bus and of having been so obviously desperate not to. An electric guitar solo blaring on the driver's radio spares me, at least, from having to talk to anyone. I duck into a green vinyl-covered seat and slide over to the window. I have sand in my socks.

Although I'm not yet ready to understand its story, the irksome sand has its own Wisconsin memories. The tiny spheres of quartz dimly recall their youth deep inside Proterozoic mountains, when they were part of a tight-knit community in granite. They attest to how erosion dismantled the mountains, how rain dissolved their neighboring minerals, how they alone survived. They remember tumbling in the surf on a tropical Cambrian beach, then lying still beneath a heavy blanket of other strata, hardening slowly into sandstone. The sand recalls how, eons later, it was excavated by rivers, then rasped by glaciers that disbanded the grains and scattered them in a meltwater diaspora. This sand has had a long and complicated journey, but this is probably its first bus ride.

Our road is near the boundary of the area served by the rural school, so the gaggle of kids who live along it are the first ones on the bus every morning and the last ones off every afternoon. We alone know the full route, where and how everyone lives: the tidy farms with bright red barns and the ones with slumping sheds and rusted machinery.

The school is in fact only a few miles down the highway, but the bus meanders back and forth along the grid of minor roads—mimicking the choreography of the tractors in the hardscrabble fields—and methodically harvests children along the way. The topography is similarly involuted; even as the bus weaves east and west, it also labors up the sandstone hillslopes and engine-brakes down again. The whole trip takes about fifty minutes.

In all those hours spent on the school bus, I assimilate the contours of the landscape, absorb the complete repertoire of early 1970s rock and roll, and observe things about society that I will not understand for years. In my mind's eye, I can still travel the route these decades later. I sit down next to my almost-eight-year-old self, watching as new kids climb on board.

First, a trio of sisters from a Mennonite family, their long, light-brown hair braided so tightly their eyes look a little squinty. They all wear handmade gingham dresses of uniform design but different pastel colors. I'm faintly envious of their interchangeability—how snug it must feel to be ensconced in a clan like that, to know exactly who you are. Then there is poor Timmy O., shuffling and unkempt. He misses a lot of school because of a congenital kidney condition that will kill him in his midtwenties. A bit farther on is the apple-cheeked kid who always brings the aroma of bacon onto the bus. It's a greasy miasma in winter when the windows are closed and the heat is on, but today it evokes a sunny farm kitchen. Again, a slight pang of jealousy, even though I am already a nascent vegetarian; ours is not a hot-breakfast family. Next, the menacing Larson brothers, before the older one lost his arm in a hunting accident. They are audacious and precocious swearers, from whom I involuntarily acquire a rich lexicon of profanities. Ours is not a cursing family.

Little by little, the bus fills up, and I take mental ethnographic notes, trying to understand what everyone else already seems to know about how to be a human being. Meanwhile, the sand reminds me of its presence, scratchy between my toes.

Sifting through the memories of childhood, it's hard to know which ones represent distinct events and which are composites. Only a few bright shards survive intact against the background of undifferentiated days. From these we attempt to create a mosaic that explains how we came to be who we are. As I imagine that school bus chugging up and down the hills, I realize how all of us—our families, the farms, and the town—like the land, were wholly shaped by sand. It defined the topography, determined the makeup of the forests, dictated where houses could be built and wells could be sunk, what crops could be grown, who got rich and who slid into debt.

Now I understand that our sandy terrain had also been a protagonist in larger, longer stories—a cultural and geological boundary zone where epic dramas played out. It lay just north of the "Driftless Area," a distinctive region of western Wisconsin and adjoining parts of Minnesota and Iowa that somehow remained ice-free during the most recent advance of the glaciers and therefore preserves an older, more rugged terrain shaped mainly by rivers. That topographic transition, in turn, coincides with what ecologists call the "tension zone," where northern pine and mixed hardwood forests give way to oak savannahs. The ribbon of land between those contrasting ecosystems acted for centuries as a natural border between the territory of the woodland peoples, the Ojibwe and Menominee to the north, and prairie tribes, the Ho-Chunk and Sauk, to the south.

It's obvious to me now that our rural community—a microcosm of midcentury America—was tainted by a legacy of racism and scarred by decades of environmental abuse, both reflecting a disregard for the history of the land prior to white settlement. The story of my town and its rocky substrate is a kind of fable, one scene in the tragedy in many acts that is now titled the Anthropocene. The stock characters are Ignorance and Avarice, the plot predictable: resource extraction briefly generates immense wealth for a small number of people but leaves an impoverished world for those who follow.

If we were ever taught in school much about the natural history of our area, or anything but the most revisionist versions of its human history, I don't remember. Any understanding of nature that I acquired came from unstructured lessons in the woods and fields around our house. My sister and I, and the few other kids who lived nearby, roamed freely across our hundred acres, picking berries, wading in the creek, or playing in the snow, according to the season, until our parents sent the signal—a blast on a vuvuzela-like horn—to come home. In this open-air classroom, we effortlessly absorbed information about the behavior of dragonflies, the habitats of mushrooms, the changing moods of the land over the arc of the year. It didn't occur to us that this was Science.

I recall a stern individual tutorial on ecosystems on a summer day when I was six or seven. My father had just built me a tree house in the low branches of a sprawling red cedar. All around its base grew bright green moss. It was so verdant and velvety that I had to have it as carpeting in my little living room. I spent hours peeling up moss and fitting the pieces tightly together on the wooden floor. By lunchtime, the floor was covered by a sumptuous emerald mat. I returned that afternoon, eager to sit luxuriously on my plush new carpet, but found to my horror that it was shriveled and browning. Although I didn't have the scientific vocabulary to describe it, I saw how stupid I had been; the moss was a symbiotic colony entirely dependent on its substrate. Ripped from its home, it had of course withered and died. I felt hot shame every time I saw the denuded area around the base of the cedar and couldn't bear to be near the tree house for the rest of the summer.

Our long sandy driveway provided lessons in hydrology and sediment transport, especially in spring, when melting snow carved small canyons and created temporary lakes. In those weeks when sledding season was over but it was too slushy for bicycling, a neighbor girl and I would volunteer as "ditchers"—self-taught civil engineers—to help drain water from the rutted road using large spoons and garden trowels. Although we did in fact accomplish this in the end, we would first impound large reservoirs of water behind earthen dams and then let them fail catastrophically, thrilled at the terrible power of the water to erode and redistribute sand in braiding patterns far downslope.

Our floods were miniature reenactments of the torrents of glacial meltwater that had left sand, pebbles, and cobbles in a nearby gravel pit where we'd look for Lake Superior agates. My father was especially good at spotting the shiny stones, with their concentric bands of red, orange, and white that resembled the growth rings of a tree. My eye would often fall on distinctive dark gray stones with short white lines in various orientations, some intersecting in a way that looked to us like calligraphic Chinese characters. I collected these as talismans, sensing that their cryptic markings must have some meaning, but I had no idea there was a science concerned with the study of such things. In any case, I never imagined that I might become a scientist of any kind; like all children, I had unconsciously absorbed gender norms at an early age.

We did learn in school that our area was once part of Henry Wadsworth Longfellow's "darksome forest" and Laura Ingalls Wilder's "Big Woods," a vast expanse of old-growth white pine, a species that thrives on sandy soil. However, I can't recall seeing any actual photographs of that monumental forest, or the people who had lived in it for centuries. Instead, we were shown heroic scenes of horse-drawn sledges piled high with gigantic logs, and burly men perched, with obvious pride, on top. We read jocular tall tales about Paul Bunyan, the gargantuan lumberjack who could fell an acre of trees with one swing of his axe.

Even as a child, I wondered whether cutting down those immense trees was a good thing, but it seemed that everyone else thought it had been. It certainly was a good thing for the late-nineteenth-century lumber barons, whose outsize fortunes created an "Empire in Pine"—and left behind the infrastructure that still defines the lives of residents more than a hundred years later. The opulent mansions of these tycoons are now museums, law offices, and multifamily apartments— reminders that no one with such wealth has ever lived in the community since.

The lake where my sister and I took swimming lessons was named for the wild rice stands that once grew on its shores, before it became the mill pond for the lumber companies. Starting in the late 1800s, a series of progressively higher dams on the river that flowed through the lake drowned the wild rice and Ojibwe burial mounds on the adjacent banks. The water was now so choked with algae fed by fertilizer run-off that no wild rice could survive anyway. After swimming lessons, our bathing suits were full of green slime that gave us itchy rashes across our stomachs and bottoms.

The public library in town was another vestige of the "pin-ery" era: a magnificent, turreted Victorian building of gargoyled sandstone. I remember many cozy winter afternoons there, surrounded by a bounty of books, sheltered within the thick stone walls that had been quarried from the same stratum, the

Dunnville sandstone, that cropped out along the banks of a creek that ran below our house. Fine and uniformly grained, this particular sandstone was ideal for the filigreed style of the Gilded Age, and for a time in the 1880s and 1890s, it was in great demand for buildings in Minneapolis, Milwaukee, and Chicago. The author of a five-hundred-page tome entitled *On the Building and Ornamental Stones of Wisconsin*, published in 1898, says admiringly of the Dunnville sandstone: "No stone takes a better face finish, or is cut and carved with better effect."[1] At the height of its fame, it was chosen for the elaborate altarpiece, sculpted by French stone masons, at Saint Thomas Church in Midtown Manhattan (adjacent today to the Museum of Modern Art). In those decades, it must have seemed that our small patch of Wisconsin, with its mighty white pines and warm golden sandstones, was building the nation.

Our library building was constructed in 1889 by one of the richest lumber magnates of all, Andrew Tainter, as a memorial to his beloved daughter Mabel, who died at nineteen. Only much later did I learn what the historical plaque at the entrance failed to mention: before Mabel (and four other children) were born to him and his wife, Bertha, Mr. Tainter, who had started as a lowly lumberjack, had had five children with an Ojibwe woman, Mary Poskin. That strategic alliance gave him access to great swaths of northern forest. As his fortunes rose, he summarily abandoned Mary, took the children into town, and hired Bertha as a governess. Two of Mary's children died before reaching adulthood, but no grand sandstone memorials were built for them.

Mabel Tainter Memorial, built from
Cambrian sandstone, Menomonie, Wisconsin

My childhood, and our town, would have been much diminished without the extraordinary wealth of the lumber
barons—they left the grand houses, the lovely sandstone library, and a manual training institute that is now a branch of
the state university system. So it took me a long time to grasp
the scale of the environmental devastation, and the intergen-
erational tragedy, of the lumbering era. Somehow, I had never
thought to use the word "deforestation" to describe it until,
as a graduate student, I was browsing at a used bookstore
in Madison and came across the *Annual Report of the United
States Geological Survey* for 1886 (the year the much-mourned
Miss Tainter died). Edited by none other than John Wesley
Powell, Civil War hero, famed explorer of the Grand Canyon,
and second director of the USGS, it included an extensive

chapter on the geology of the Lake Superior region and, in particular, its iron- and copper-bearing rocks.[2] Although the tome seemed an extravagance then at twelve dollars, I bought it.

The text is illustrated with a dozen detailed lithographs of outcrops that are still classic stops for student field trips in the Baraboo and Gogebic Ranges, Wisconsin's erstwhile mountains. Many of these sites are now within state park or national forest boundaries. But in the images in the USGS report, the landscape is completely naked, with the pathetic, ragged look of just-shorn sheep. Not a single tree has been left standing. I admire the careful work of the nineteenth-century geologists who worked in the Upper Great Lakes region; their maps and observations have survived the test of time, enduring the erosive power of scientific paradigm shifts. Yet these keen-eyed scientists make no mention of the utter denudation of the landscape, which would have happened just a few years before they wrote their reports. Maybe they thought it was a boon. Certainly, it would have been easier to map the bedrock for Mr. Powell without all those bothersome trees.

What did the average Wisconsinite think of the shocking deforestation at the time? Most white settlers accepted the ravaged landscape as the price of progress, a patriotic imperative; the land *had* to be cleared for agriculture. The dense old, wild forests—not yet called "ecosystems"—did not merit conservation, or even documentation. The trees were worthy of photography only after they had been brought down. Even Wisconsin's now-treasured National Forest, the Chequamegon-Nicolet, is just regrowth on the cutover, a tragically impoverished version of the Big Woods. One patch of the great forest does survive on the reservation of the Menominee Nation not far from Green Bay. As a result of the tribe's long-term practice of selective harvesting, the reservation is a verdant rectangle visible from space, a lush oasis of trees amid a desert of fields and scrubland. But the Menominee forest is just a tiny remnant of what was here before the lumbering era, and most people who live in Wisconsin today have no idea of what has been lost.

I've glimpsed ghosts of the Big Woods and tried, as an academic, to conjure that lost world back into existence. In the late 1990s, a company in northern Wisconsin began to salvage from the floor of Lake Superior logs that had sunk more than a century before, when lumber camps floated timbers down rivers into Chequamegon Bay to be stockpiled until they were sawn. The end of each log was stamped with a unique symbol representing the camp it came from and the company that owned it. Fully 10 percent of all the logs that were cut became waterlogged and foundered to the frigid lake bottom, where they lay for a hundred years. The cold, low-oxygen conditions preserved the cellular structure of the wood, especially of deciduous species like oak, maple, and beech. Late-twentieth-century entrepreneurs saw an opportunity to benefit from the wastefulness of their nineteenth-century predecessors, selling the wood for veneer, artisanal furniture, and musical instruments.

My father, a specialist in wood technology, was hired as a consultant to help advise how the waterlogged wood could be dried without cracking. When I saw some of the logs, it occurred to me that their rings could be archives of climate information from the centuries before the forest was annihilated. The company generously gave me end-cuts from two dozen logs, most of them with the legible stamps, making it possible to determine where and when these were cut. I then enlisted a group of enthusiastic students to begin the tedious task of counting and measuring the rings.

We quickly found that this could not be done by eye or even a 10X hand lens; most of the rings were so narrow that we had to set up a special microscope rig to measure them. These trees, all hardwoods, had grown exceptionally slowly. The cross sections were less than two feet in diameter, but the trees were 250 to 300 years old when they were cut. In many of the specimens, there were intervals of eight to ten years when rings were wider than average, but to our frustration, the times of more rapid growth had occurred in different calendar years in different trees. In the end, we realized that the growth of these hardwood species had been limited not by precipitation or harsh winters but by access to light. They had been eking out a living in perpetual twilight beneath the towering canopy of white pine. It was only when an old pine fell that sunshine could reach the understory, and for a few glorious years these light-starved trees added fat rings. This was not the scientific result we had hoped for—we didn't learn much about historical precipitation patterns—but we felt as if the forest primeval had spoken to us from beyond the grave.

For the Indigenous people of the region, the Ojibwe and Menominee, whose cultures were wholly integrated with the plants, animals, and waters of the woodlands, the deforestation was apocalyptic. Most tribal reservation boundaries had been established in Wisconsin by the Treaty of 1854, confining the tribes that once had occupied the entire Upper Great Lakes region to a few scraps of what was considered undesirable land. But when it became clear that reservation land hosted valuable lumber resources, the US government suddenly desired it.

In the infamous 1872 case *United States v. Cook*, the Supreme Court ruled unanimously that the tribes did not in fact own the land on their reservations and could not sell any timber from it—unless the trees had been cut for the purpose of establishing farmsteads. Any trees cut for other reasons were deemed the property of the US government—or in reality, corrupt local officials allied with the lumber companies.[3] The farmstead exception was based on the patronizing and illogical notion that the Ojibwe and Menominee would become "civilized" by turning to farming—in a region with short growing seasons and shallow, sandy soils.

Even the area around our town, well to the south of the reservations, was at the northern edge of productive farm country. And, in a tragic irony, the clear-cutting in the late nineteenth century forever changed the hydrology of the landscape, making agriculture even less viable. Where the thirsty roots of towering white pines had once soaked up and helped retain rainwater in the soil, there was now only scrub, pasture, and cropland. Rainwater and snowmelt coursed over the landscape unimpeded. Erosion was ferocious, and the already marginal sandy soils were quickly washed away. Floods became more frequent and extreme, and sediment load in rivers increased by 500 percent, leaving a measurable stratigraphic record in floodplains and lakes.[4] Even if we would like to delete from historical records the shameful facts of the lumbering era, they are indelibly recorded by the land itself.

Soil conservation became a priority in Wisconsin and the rest of the country in the 1930s, as it grew obvious that agricultural practices were dramatically accelerating the natural, inevitable geologic process of erosion—an early sign of the looming Anthropocene. Like almost all country kids, I was in 4-H, the youth-mentoring organization founded to train future farmers, and we learned a bit about preventing soil loss from US Department of Agriculture filmstrips and brochures. Every 4-H member was also given a bundle of fifty white pine saplings to plant each year, which we moaned about at the time, but I now see it as an enlightened lesson in environmental citizenship.

In other ways, 4-H was not so enlightened. All members needed to sign up for at least three "projects"—dairy, poultry, sheep, hogs, gardening, woodworking, knitting, etc.—that would yield animals, vegetables, or crafts to exhibit at the county fair. I wasn't a farm kid, so raising livestock wasn't an option, and the choices were further constrained by unwritten but closely observed gender rules. One year, out of naïveté, I signed up for woodworking. My father was a master cabinetmaker with a well-equipped workshop, and I already used his tools to make furniture for my dollhouse. As it turned out, I acquired few new woodworking skills from the meetings of the 4-H group but did learn to make myself as small and invisible as possible in male company, like Jane Goodall crouching in the forest, documenting the behavior of chimps. My observations would prove invaluable years later in decoding male conduct in physics classes, faculty meetings, and remote field camps.

I was also in the gender-appropriate 4-H sewing project, which culminated each year in the Dress Revue, where members modeled garments they had made. Resisting my mother's advice, I chose a pattern with complicated bodice pleats and frilled cuffs, which I cried and labored over for weeks. When at last the dress was done, I felt sure that my technical virtuosity would be recognized by the judges. On the morning of the Revue, I got a ride into town with the older neighbor girls. They hadn't managed to finish their frocks and had resorted to stapling the hems. Sitting in the backseat, I smiled and thought smugly about my masterpiece.

When we arrived at the Revue, however, I realized I hadn't given much thought to the modeling part—how *I* would be under as much scrutiny as my dress. When my turn came to walk and twirl before the female judges, I cringed and cowered, attempting to hide my flat chest, bitten nails, and ungainly feet. The judges barely looked at the pintucks and button-holes I had toiled over. The beautiful neighbor girls strode confidently up in their stapled dresses and took blue ribbons. I learned something that day about how women can judge each other and resolved not to be part of a system where appearance mattered more than substance. Although I can still feel the sting of humiliation, being told early on that I was not a pretty ornament spared me from years of pinning my self-worth to external standards of beauty.

In the early 1970s, the country was being rocked by social upheaval, but traditional norms prevailed in our community. By the time I was eight or nine, I began to be aware that our family was culturally anomalous. When *Ms.* magazine was first published in 1972, my mother immediately subscribed, and I studied every issue with scholarly intensity. We lived in a house that my father had built—was still building—in the woods. For a year after we moved in, there was no staircase, only a ladder between floors, but there were bookshelves from the start. I remember visiting other people's houses where only family photos lined the walls and wondering where they kept all their books. My parents were generous hosts; dinner guests included other faculty members from the state college;

homesick international students from Nigeria and Trinidad; and back-to-the-land types from the Twin Cities who had, with endearing optimism that outstripped their technical skills, bought old, decrepit farmsteads and needed my father's advice about cracked foundations and sagging beams.

My parents grew up in the same small town in northwestern Minnesota and were the first in their families to attend university. My father's parents were stolid, taciturn Norwegian Americans who ran a dry-cleaning business. Whenever we went into their shop, the chemical fumes burned my throat and stung my eyes. I couldn't imagine my grandparents breathing that air day after day for years. My mother came from a more tumultuous family background. My maternal grandparents had a fractious marriage, and my grandmother was desperately unhappy. When her sister's husband was killed in a farm accident in 1941, the two women left Minnesota abruptly to work in the wartime shipyards in Seattle— leaving my mother and her sister, then four and eight, to live with my hard-drinking grandfather. Although I have known of this family story for years, it still shocks and troubles me; I don't know how to interpret it or dispel its power. As a daughter, I know how it cast a permanent shadow on my mother, a pervasive sense of dread—beyond the baseline Scandinavian level—that I absorbed early on and have no doubt unintentionally passed on to my own children. As a mother, I can't imagine abandoning my children, but having known the suffocating feeling of a bad marriage, I can also understand the animal survival instinct to escape.

After my parents graduated from a nearby teachers college—in a town with an especially large effigy of Paul Bunyan—my mother taught English at the high school in a town on the edge of a large Ojibwe reservation. This experience, and my mother's visceral empathy for motherless children, led my parents to adopt my sister, born in Chicago to a teenage Ojibwe mother who had given her up at birth. My sister had been in five different foster homes by the time she joined our family at age nineteen months in the tumultuous summer of 1968 (a decade before the Indian Child Welfare Act, which seeks to keep Indigenous children in their communities). I was almost six at the time, and from that point on, through my sister's eyes, I became aware of ubiquitous, casual racism and its corrosive effects. Taunting at school caused her to be always on guard, bracing for the next jab. As awkward as I was in social settings, I saw that she did not have the luxury of slipping into invisibility, an option that I relied on and took for granted.

Our family had connections to a remarkably wide range of subcultures, which allowed my research on humans to extend to settings beyond the school bus. We attended powwows held by local tribes, eating fry bread, joining the shuffling dance circles, and falling under the spell of the drums. We were members of the Norwegian Lutheran Church (there were also Swedish, Danish, and German ones in town), and though its liturgy failed to stir any spiritual response in me, its social network was an important scaffolding for our lives.

My parents had close friends who lived in the Minneapolis area, and their daughter and I saw each other as city mouse/country mouse counterparts. She was fascinated with the "party" telephone line we shared with three other nearby households, and I taught her the art of picking up the receiver very slowly in order to eavesdrop on conversations without being detected. When we visited them in the Twin Cities, we would sometimes stop after hours at the upscale furniture store where her father worked, and I loved moving from one pseudoroom to the next, each furnished with books that created an eerie verisimilitude of an actual living space. It felt like trying on other people's lifestyles, testing hypothetical futures.

Most of our family's furnishings, however, came from farm auctions, which provided especially rich opportunities for ethnographic observations. My parents had a good eye for spotting well-made tables and wardrobes waiting to be liberated from bad coats of varnish. Almost every item of our furniture had had some earlier, secret life before coming into our house. Sometimes I felt as if I could hear the chairs reminiscing about long-ago gatherings, the tables recalling hopes and aspirations that never materialized.

We children were not insulated from tragedies in the community. One spring evening when I was in the second grade, a schoolmate was riding his bike at dusk near his family's farm, set in a beautiful spot at the base of a sandstone hill that forces the road into a tight bend. A milk truck driver coming home after a long day rounded the curve, and, perhaps blinded by the low sun, struck the boy. One of the sharpest shards of memory from my childhood is the day our class walked silently behind our teacher from the school to the country church where the boy's waxlike body lay in a white satin–lined coffin. Overcome by the sickly, euphemistic scent of flowers, I lurched for the door, desperate for sunlight and open air. All these years later, with three mortal boys of my own, the memory of that first encounter with death still pierces me.

By the time of my childhood in the early 1970s, it was becoming difficult to make a reasonable living on a small family farm. The hilly terrain made it impossible to create fields that could accommodate new, larger equipment for more efficient plowing, planting, and harvesting. Farms with only a few dozen cows could no longer compete with the growing number of highly mechanized megadairies. This was the beginning of a long, slow, downward slide for our area—and much of rural America—that has now hardened into bitterness. But at that time, there was just a spreading sense of exhaustion.

And the land itself was tired. With each successive year, the sandy soils required more and more nitrogen fertilizer to yield a reasonable harvest. Whenever it rained, some of those nitrogen compounds were flushed down into the porous sandstone bedrock, from which most households drew their drinking water. At a growing number of farmsteads, tests of well water revealed high levels of nitrates, compounds that strip oxygen from hemoglobin in the blood. Many people viewed this with a sense of resignation; it was the price of progress, the reality of modern agriculture. But I remember the hushed whispers of horror among the neighbors when one of our former babysitters, who had married a hardworking young farmer the previous year, gave birth to a "blue baby" who died within hours, poisoned in utero by nitrate-contaminated groundwater. The pain of the tragedy was amplified by the unspoken sense of collective culpability and the insidious, subterranean nature of the cause.

The science of groundwater, or hydrogeology, is now a sophisticated, highly quantitative subdiscipline, but even today it retains in the public imagination a whiff of the occult. People who would scoff at palm reading or crystal healing will still hire dowsers or water witches with divining rods to determine where to site a well. This could be dismissed as laughable if widespread ignorance about groundwater didn't endanger public health.

Hydrogeologists, who study the secret world of groundwater, look at rocks through a different lens than do other geologists. A sedimentologist, for example, examines a sequence of rock layers and sees the ancient environments in which they were deposited—in the same way that a historian may study the street plan of a city and find clues as to how it evolved over time. A hydrogeologist, in contrast, is more like a transport engineer, less interested in history than in how efficiently particular rock layers can transmit groundwater "traffic." On a watery planet like Earth, almost all rocks, regardless of their origins, eventually interact with groundwater. Some rocks discover, hundreds of millions of years after they formed, that they are very good at storing and conducting groundwater and take on new careers as aquifers.

The Cambrian sandstones of Wisconsin, those golden beach sands favored by Gilded Age stonecutters, constitute the most important aquifers in the state. The pounding surf of 500 million years ago rounded the quartz sand—the stubborn residuum of ancient mountains to the north—to near-perfect spheres and sorted them by size. Like marbles in a jar, the grains don't nest tightly together, leaving plenty of room for water in between. Wherever it is exposed at the surface, the Cambrian sandstone, always a generous host, thus welcomes in rainwater and snowmelt and transmits it speedily downgradient.

My first inkling of the hidden world of groundwater came from the seeps and springs in the sandstone banks along the creek below our house. In the spring and summer, water trickled out from a layer in which the sand grains were cemented together by iron, leaving rusty red streaks down the cliffs and providing a habitat for peppery watercress. In the winter, great cascading icicles would form on the banks, some like stately architectural colonnades, others suggesting the fangs of monstrous creatures.

The sandstone's open-access policy for incoming rainwater is both a blessing and a liability. It keeps the aquifer reliably refilled but also allows entry to any toxins and pathogens that happen to be hitching a ride, like pesticides or disease-causing bacteria. And once those have gained admittance, they tend to stay, often sticking to rock surfaces underground and thereby maximizing the amount of water they can contaminate. Even if we would like to absolve ourselves of environmental crimes, the groundwater will remember and testify against us, exacting terrible penalties.

I look back with astonishment at how far our understanding of geologic processes has come in the last few decades—and what wasn't known even as recently as my schooldays. Fields like hydrogeology were hampered by the lack of computational capacity: the basic equations governing groundwater flow were known, but it was virtually impossible to do the calculations by hand in a rigorous enough way to create predictive three-dimensional models. In other subdisciplines, the problem was a lack of unifying conceptual frameworks for explaining, for example, how it could be that Wisconsin once had imposing mountains, or expansive tropical beaches.

It's a little embarrassing for geologists to acknowledge that the age of the Earth was not determined until 1955, and, even worse, that the origin of granite—the very foundation of the continents—was still in dispute at that time. Plate tectonics, which is the way the solid Earth works—how mountains grow, why earthquakes and volcanoes occur where they do—wasn't figured out until the late 1960s. The crucial role of Life in modulating the chemistry of the oceans and atmosphere only began to be appreciated in the 1970s. High-resolution records of past climate weren't available until the 1980s, and the idea that human activities could rival natural geologic processes as agents of global change didn't gain widespread acceptance until the 1990s.

Stratigraphy—the study of layered rock sequences—was perhaps the most mature branch of geology at the time of my childhood. Sedimentary rocks like sandstone are the most accessible geologic "texts"—we can easily imagine how they formed by looking at modern settings where similar deposits are accumulating today, and we intuitively grasp that a layered sequence records the passage of time. Not surprisingly, then, stratigraphy was the starting point for the science of geology, which began to define itself as a discipline in the early 1800s. A British canal digger named William Smith, who spent his days blasting through hills of stratified rocks, made a brilliant discovery: particular layers, though missing from eroded valleys, could be traced across the width of England based on the consistent sequence of distinctive fossils within them. Using this astute observation, Smith not only drafted the first modern geologic map but also provided a basis for global correlation of rock units, the key to building the geologic time scale. That task would preoccupy many subsequent generations of geologists.

From the earliest days of geology, understanding regional stratigraphy was not merely a scientific endeavor; being able to predict where coal seams might occur in the subsurface provided strong economic motivation for studying sedimentary sequences. The field of stratigraphy got an even bigger boost in the early twentieth century as petroleum exploration evolved from wildcatting to a more systematic enterprise. For decades, the leading journal on stratigraphic research was the *American Association of Petroleum Geologists Bulletin.*

By the 1960s, legions of field geologists had mapped, named, and described sedimentary strata across the North American "craton," or stable interior, where the rocks had not been crumpled or faulted by tectonic forces around the edges. Most of these stratigraphers focused on interpreting the history of a single region, particularly the southwestern United States, where canyon lands served as open books that told of waxing and waning seas, vanished landscapes, and long periods of erosion.

But a few big thinkers, notably Laurence Sloss of Northwestern University—in the heart of the craton—searched for continent-wide patterns in rock sequences and what these might reveal about long-term, global changes in sea level. The conceptual difficulty of this undertaking, before computers facilitated spatial visualization, can hardly be overstated; identifying such patterns required depicting rock sequences at the continental scale in four dimensions—i.e., envisioning their forms not only in space but also across time.

In a 1963 paper that is still assigned reading for students of stratigraphy,[5] Sloss proposed that six great sedimentary sequences had been deposited across North America over the last 500 million years. Each is bounded above and below by major erosional surfaces called "unconformities," recording six major cycles of sea level rise or "transgression," and fall or "regression." (When, in an undergraduate course, I first learned of the "Sloss cycles," I heard the phrase as "slosh cycles," which evoked an image of the world's oceans swilling back and forth in slow motion over tens of millions of years.) These cycles reveal a restless global ocean that waxes and wanes, retreating in ice ages and advancing in warmer times, in an endless on-again, off-again relationship with the continents. Sloss's work was a conceptual revolution, the seed for the new field of "sequence stratigraphy," which quickly spawned graduate programs and professional conferences—and became the paradigm that shaped all subsequent exploration by the world's oil companies.

Sloss named his six sea-level cycles for Native American nations that occupied the places where the associated rocks are most extensively exposed. From youngest to oldest the cycles are: Tejas, Zuni, Absaroka, Kaskaskia, Tippecanoe, and Sauk—the earliest being named for the Sauk people of Wisconsin, because the rocks that most clearly document the beginning of this continent-wide inundation are the Cambrian sandstones of Wisconsin. Sloss himself, in Evanston, probably drank daily from the Sauk Sequence sandstone aquifer, which slants downward into the subsurface toward Lake Michigan and is tapped by municipal wells in many Chicago-area communities. Wisconsin's Sauk sandstones have "cousins" across the continent, including the Tapeats Sandstone in the Grand Canyon, the Flathead in the northern Rockies, and the Potsdam in upstate New York.

Among all the Sloss cycles, the Sauk sequence is the most clearly defined. Its base is the Great Unconformity, a profound erosional surface most famously exposed in the Grand Canyon—the discontinuity between the ancient metamorphic rocks of the lower Canyon and the younger, horizontally stratified rocks above. It marks the boundary in time between an Earth with only microbial life and one whose oceans teemed with the first animals. Each of the Sloss sequences begins with an unconformity overlain by sandstones that record the migration of beaches across the landscape, but the Sauk sandstone, as Victorian stone carvers noted, is qualitatively distinct. Sedimentologists describe it as "supermature"—processed and refined through so many cycles of chemical weathering and physical abrasion that only the most stable and resistant residuum of perfectly spherical quartz grains remains.

Both in outcrops and under the microscope, the Sauk sandstones—also called quartz arenites (from the Latin word for sand, *arena*)—are things of beauty, causing geologists to abandon the normal constraints of scientific writing and wax lyrical. In roadcuts where highways slice through sandstone hills, the fresh rock ranges from the creamy color of vanilla pudding through a spectrum from honey to caramel. At high magnification, the round, smooth grains resemble white beluga caviar. One of my own mentors, the eminent sedimentologist Robert Dott at the University of Wisconsin, spent much of his life studying the Sauk sequence sands. (His name was apt for someone immersed in the pointillist world of sand.) Late in his career, in an article published in a normally somber professional journal, Professor Dott allowed himself to express his true feeling about these rocks: "I have come to regard supermature quartz arenites as nature's finest distillate— almost as remarkable as a pure single malt Scotch whiskey."[6]

Just as the best distilleries keep their methods of production secret, the supermature Sauk sandstones hold on to their mysteries. Quartz is not the most abundant mineral in the

Earth's crust (the large group of minerals called feldspars are)
but is simply what survives when everything else is worn away.
The immense volume of quartz in the continent-wide Cambrian sandstones is astonishing when one realizes that five
or six times that volume of continental source rock of typical
granitic composition would have had to go into the "hopper"
to produce it.

To account for the existence of such extraordinary rocks,
Professor Dott boldly challenged one of the foundational
concepts in geology, the principle of "uniformitarianism"—
the idea that geologic processes have remained the same over
time. Professor Dott argued, radically, that the Sauk sandstones are unique, the product of a never-to-be repeated moment in Earth's history. In particular, the Sauk transgression
followed a protracted period of low sea level and continental
erosion of a duration that has not occurred since. Moreover,
the continents at the time were completely unvegetated; rivers
would have run unchannelized in broad, braided floodplains
across much of the land surface, and unchecked winds would
have assisted in winnowing away silt and clay. Then as seas
rose, crashing surf performed one last stage of sifting and
sorting, yielding an extraordinarily refined product. By late
Ordovician time, about 450 million years ago, when algae had
evolved into plants that were establishing new ecosystems on
land, it just wasn't possible to distill a fine sandstone the way
it was done in the Cambrian. But this is not to lament the rise
of vegetation! How fortunate we are to enjoy both a verdant
Earth and the ivory sands of a long-vanished world.

The very homogeneity of the supermature Cambrian
sandstones is another reason why they have remained enigmatic. Quartz is lovely, translucent, and pure but, frankly, a

little inarticulate. And one quartz grain is much like another; in aggregate, they have stories to tell, but no single grain recalls precisely where it was born or how old it is. Luckily, one out of every ten thousand grains in even the ultrarefined Sauk sandstones has a better memory. Hidden in a haystack of quartz are rare needles of zircon, a mineral both tougher and more eloquent than quartz.

Unlike quartz, which does not keep track of its age, zircon is ideally suited for isotopic dating. At the time of crystallization, it allows some stowaway uranium into the berths normally occupied by the element zirconium in its three-dimensional molecular structure. Over time, the radioactive uranium decays to distinctive isotopes of lead, and measuring the lead-to-uranium ratio in a zircon crystal allows it to be dated with high precision. Like trees that have growth spurts when they get new access to light, zircon crystals can accrete new concentric layers when reheated in tectonic events that create volcanoes or raise mountains, laying down growth bands that resemble miniature tree rings. A single zircon crystal can contain the tectonic history of a whole continent. Quartz grains in the Sauk sandstones of Wisconsin gesture vaguely toward ancient mountains in the north, but the zircon grains remember specific events in their construction.

Quartz is the blurred memories of childhood, the indistinguishable days spent gazing out the window of a school bus. Zircon is a sharp, bright shard from a distinct moment in the past that has somehow survived intact.

Forty years after Sloss's insights about the Sauk and other sedimentary sequences revolutionized petroleum exploration, Big Oil began to take a new kind of interest in Wisconsin's iconic sandstones. Although the rocks do not themselves contain oil or gas, the advent of hydraulic fracturing or fracking—i.e., cracking open impermeable shale formations for hydrocarbons—turned the golden sandstones into golden geese.

Fracking involves the injection of water and chemicals at high pressure into the subsurface to create networks of tiny fractures that allow extraction of fossil fuels. When the fluids are pumped out, these cracks would collapse if they were not somehow propped open. As it happens, all that vigorous multicycle processing of the Sauk sequence yielded the perfect "proppants" to fill those cracks: exceptionally round, size-sorted grains of pure quartz sand.

Struggling farmers in the sandstone terrain of western Wisconsin faced a Faustian reality: the bedrock beneath their fields was worth more than years of hard-won harvests. Many of them chose to sell their land for sand mining. In some counties, the topography changed almost overnight; the rolling hills were deleted with ruthless efficiency as frac sand operations stripped away vegetation and gouged out the golden bedrock, then rabbled on, just as the lumber barons had done a century earlier. An observer making long-term ethnographic observations might discern a pattern: First, Indigenous people of the forest were forced off the land so the sand-loving white pines could be cut, then the sandy soil was lost to erosion, the sandstone aquifer was tainted, and when it seemed that nothing else could be subtracted or diminished further, the hills themselves were hauled away. The history of this place has always been dictated by sand.

Countless grains of Wisconsin's Cambrian beach sands, finer than single malt whiskey, now lie wedged in tiny fractures within sulfurous shales deep beneath Pennsylvania, Texas, and

North Dakota. I imagine their claustrophobia at being trapped underground again—and their confusion in finding themselves in completely unfamiliar Sloss sequences. Better to be traveling up and down their native hills on a school bus, in the sock of a girl who could one day share their story.

2

BASALT

Three university vans—the old fifteen-passenger, rollover-prone kind—are hurtling along Highway 61 on the North Shore of Lake Superior on a chilly Saturday in May. I'm in the third vehicle, sharing the last row of seats with tents and sleeping bags jammed up to the ceiling. We'll be camping at the end of the Gunflint Trail, on the east side of the Boundary Waters, close to the Canadian border. This is my first multiday field trip as a geology major, and I'm just beginning to develop a sense for the norms of this particular subculture. I hear the teaching assistant, a bearish, bearded PhD candidate who is driving with alarming inattention, say that snow is in the forecast, and I curse myself for not bringing a better coat. It didn't even occur to me this morning when we left the balmy spring weather in Minneapolis.

We're well past Duluth, and the few businesses along this stretch of road aren't open yet; the summer tourist season is still several weeks away. Our professor announces on the staticky walkie-talkie that we'll make a rest stop, so the vans pull over to the shoulder of the highway, and the thirty of us unfurl ourselves into the bracing air. It's 1981, the oil industry is booming, and geology departments everywhere are bursting with students. Geology majors could land high-salary jobs in

Texas and Louisiana in oil and gas exploration with even a partly completed bachelor's degree.

The men in our group quickly relieve themselves by the side of the road without a hint of self-consciousness, while I and the few other women in the class glance at each other uncertainly. There's no spot that will be out of view of the others, especially since the guys are now fanning out and heading down to the pebble beach. In wordless consensus, we all decide to hope for a bathroom somewhere soon and join the professor and our male classmates by the shore.

Our small miseries are soon forgotten as the Big Lake starts working its charms on us. Its vast, wild solemnity compels our attention. Lake Superior is the largest in the world by area, containing 10 percent of all the fresh water on Earth's surface. It creates its own weather patterns, and its tempests have swallowed at least five hundred ships whole. Today, though, the water is fairly calm. We can just make out the low silhouette of Isle Royale on the horizon to the northeast, but the far side of the lake, more than two hundred miles away, is hidden by the curvature of the planet.

The water is brain-freezingly cold, and with air temperatures in the forties, no one is tempted even to dip a toe in. Yet we all feel some visceral need to interact with the magnificent, mysterious expanse of water. The guys start gathering flat stones for skipping, falling comfortably into habits from boyhood. I walk along picking up unusual-looking rocks, knowing that there could be quite a variety here, gathered from a wide swath of Ontario by glacial ice. Soon, I've amassed more stones than I can hold and am certain that some are rare finds. One is almost sensuously smooth, with a matte finish of rich forest green; one is like a rigid, rusty sponge, full of spherical holes; a third is somber gray but gaudily polka-dotted with orange crystals in starburst patterns.

Then I spot a familiar friend from childhood—one of the "Chinese calligraphy" rocks we often found in the gravel pit at home, with needlelike crystals that seemed to encode messages. Perhaps now, at last, I can learn what their inscriptions mean. Our professor is looking out toward the lake, apparently in no hurry to move us on to the next stop. Back on campus, I'm generally too timid to ask questions in class and could not imagine a one-on-one conversation with a faculty member during office hours. But here in the presence of this Great Lake, whose immensity flattens human hierarchies, I am bolder. I show the cobbles to the professor, a soft-spoken man whose meandering lectures both reflect and obscure his deep knowledge of Minnesota's geology. He looks quickly at my collection and says, "This one's basalt, so are those two, and same for this porphyritic one with the long crystals of plagioclase in a fine matrix. They're all local basalts—haven't traveled far."

All basalt! I receive this news with a strange mix of emotions. First, embarrassment; this is an intermediate-level course in igneous petrology, the study of rocks formed from magma, and even a student in an introductory class would know that basalt—the volcanic rock of Hawaii and Iceland—is among the most common rock types on Earth. *How stupid of me*. It's as if I've missed a basic vocabulary word on a language quiz. Next comes a burr of irritation, perhaps akin to what people learning English must feel about wildly inconsistent spelling and pronunciation rules: *None of these look like the basalt specimens in the classroom—How can one kind of rock have so many*

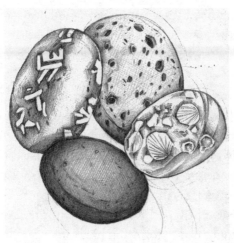

Cobbles of basalt from the North Shore of Lake Superior

different appearances? Then, a shadow of self-doubt: *Maybe I'm not cut out for geology; I just don't have an eye for rocks.*

Yet once my wounded pride has recovered, I feel a nascent shift in perspective. The calligraphic—porphyritic—rocks we treasured as children were more than mere curios. They were legible records of a vanished world where black lavas oozed from subterranean fractures. The "characters" in the rock were clumps of slender crystals that had formed in a subterranean magma chamber before being swept to the surface in an eruption of liquid rock. I had sensed that these stones had secret wisdom to share, not realizing that they were remnants of a great volcanic empire. Now, it was suddenly obvious to me that all of human civilization was perched on top of ancient geologic kingdoms.

We never stopped anywhere with an indoor bathroom that day, and it did snow in the night. But by the end of the trip, in the presence of that immense, implacable lake, the group had been unified by deepening engagement with a story that was much larger than any of us. We were beginning to hear the rocks speaking.

Can we ever revisit the past without seeing it through the lens of what happened later? Can we truly understand a bygone time in its own right, when other outcomes were still possible? Even when studying rocks, it's hard to avoid the bias of presentism. The Lake Superior basalts formed a billion years ago, on an Earth when life was still entirely microbial. Trilobites were still 450 million years in the future; another supercontinent had to form and break up again before Pangea would even begin to take shape; there was no reason to imagine that such things as dinosaurs or primates would inherit the Earth. None of these things—though now literally set in stone—were then preordained; given the happenstances of planetary and biological evolution, Earth's story could have unfolded quite differently.

My path into geology felt anything but predestined. I had always been a good student, a bit of a smarty-pants, I admit, facile with words and pretty good at math. I entered college intending to major in languages or linguistics—with some naive idea of becoming a translator—and registered for an introductory geology course in my first term so I could check off the lab science box on the list of requirements. The course was taught in the old-fashioned way, marching systematically through encyclopedic lists of minerals, rocks, and landforms.

At first, as an outsider from the humanities, I simply reveled in the idiosyncratic lexicon of geology. The names for landforms and rocks were a hodgepodge of references to mythology, anachronisms from alchemy, and imports from many languages. "Basalt," for example, is a medieval corruption of a Greek word, *basanites*, which in turn was a variant of *babas*, an ancient Egyptian term for "touchstone"—any dark, fine-grained rock that assayers used to test whether a metal was truly gold. The real stuff would leave a yellow streak on the black touchstone.

But after a few weeks, I began to understand the stones themselves as records of far more ancient worlds: each a text to be translated, a portal into the hermetic inner life of Earth. Of course, none of my professors spoke about rocks this way; most of them intoned lectures without inflection, seemingly inured to the strangeness of the planet—its self-renewing tectonic habits, its ceaseless repurposing of primordial ingredients, its literary impulse to record its own history. This was science; metaphor and mysticism were not allowed. But I annotated my notebooks with marginal comments and exclamation points at the implications of what we were learning: The solid Earth was animate—cyborg-like—truly bizarre.

I look back with amazement that my sixteen-year-old self recognized something in geology that would continue to engage me intellectually and ground me personally in the decades to come. Many misguided life decisions would follow, but at least I got that one right.

Becoming a geologist requires an unusual combination of conceptual skills. One is a facility for zooming in and out of scales in space and time—to read any single rock specimen or outcrop as a signifier of a larger geologic tableau, and to travel in the mind's eye with a rock across eons, from its formation to its intersection with our temporal plane. These days, it is also a technologically advanced science in which fieldwork is supplemented by investigative approaches including instrumental analyses, laboratory experiments, computer simulations, and real-time monitoring of the oceans, atmosphere, and solid Earth.

Yet even today, accurately identifying rocks, minerals, fossils, landforms, faults, and other natural features remains essential to the practice of geology. Trying to do geology without knowing such things by name is like studying a language without learning any actual words. The names themselves are, of course, human constructs, but the act of naming requires making distinctions that sharpen the powers of observation. Like other natural sciences, geology *had* to start with taxonomy; entities cannot be investigated until their existence is recognized. Given the sheer profusion of things to inventory on Earth—the accumulated artifacts of billions of years—it's not surprising that the process preoccupied geologists for the entire nineteenth century. Unfortunately, the field has not been able to shake a musty reputation as a descriptive science concerned mainly with collecting, naming, and classifying rocks. Today, when anything can be effortlessly googled, taxonomy seems especially old-fashioned.

But categorizing things is neither as tidy as its practitioners would like to think, nor as trivial as its detractors suggest. In the hierarchy of sciences, fields that involve description and classification of wild nature have long had lower status than those built on controlled experimentation. In a much-repeated slight—which, ironically, is itself taxonomic—early twentieth-century physicist and Nobel laureate Ernest Rutherford is reputed to have said: "All science is either physics or stamp-collecting."[1]

Taxonomy is comforting because it creates a sense of control and finitude in a chaotic and open-ended world. I understand the attraction of categorization; for a time, one of my favorite possessions as a child was a gallon-size jar of old buttons that was a lagniappe—a bonus item—thrown in with a box of other stuff my parents bought at a farm auction. I loved spilling the buttons onto the floor and sorting them by size, color, or design. But I found there was never a single system by which all the buttons could be satisfactorily classified. The categories were either too large or too small, and some buttons could be placed in several groups: upholstered, green, square. The exercise was always inconclusive; the buttons would go back into the jar until my next attempt. An early sense that this might be an odd pastime came at a 4-H summer camp when I was eight or nine. We were sitting around a bonfire the first night and asked to introduce ourselves by telling the group about something that we liked to do. The rest of that miserable week, I was the weird button-sorting girl.

My early experience with category-defying buttons was good preparation for the untidiness of geologic classifications. Try as we might, geologists have never been able to impose comprehensive taxonomies on all the components of unruly Earth. As with the buttons, the problem is determining which attributes deserve the most attention; rocks, for example, might be classified by color if one is making pigments, or tendency to cleave if one is tiling a roof, or sculptability if one is Michelangelo—but none of these may be particularly relevant to understanding their origins.

Geology's obsession with nomenclature has sometimes led the field astray, because it puts the focus on objects rather than events, depicting rocks as inert matter and not the shape-shifting time travelers they are. Knowing and naming rocks may not seem too different from identifying trees or birds—but consider that in most places, rocks are not found in the habitats in which they formed. A forester in Minnesota has no chance of encountering a stand of mangroves, and an ornithologist will not see a flamingo on the shores of Lake Superior, but a geologist often encounters rock "species" formed by processes completely unlike those that occur in their modern settings. There are no volcanoes in Minnesota today, but there is a lot of volcanic bedrock. So a Minnesota geologist needs to become familiar with the habits and customs of volcanoes—as well as all the other agents of rock production. This makes the first years of geologic training especially challenging, but it also means that geologists learn a global lingua franca that allows us to read the stories encoded in rocks anywhere in the world, and even on other planets.

Igneous rocks like basalt are defined by their constituent minerals (in the case of basalt, about half calcic feldspar and half clinopyroxene), because this reflects their magmatic sources. But as I learned on my first field trip, an igneous rock of any given composition can have a wide range of outward guises that reflect specific processes that occurred as it solidified from magma to rock, and in its subsequent experiences over long spans of geologic time. In other words, every rock encodes its own autobiography.

A basaltic lava that is full of magmatic gases (mainly carbon dioxide and water vapor) will develop a frothy crust on its surface as gas bubbles attempt to escape, creating a rock full of holes, or "vugs," like the spongey cobble I collected on the Lake Superior shore. Geologists—who brandish an absurdly large lexicon of adjectives—call this "vesicular" basalt. A less gassy lava that cools quickly and has no time to grow large crystals will quench to a fine-grained mass like my exquisitely smooth stone—an "aphanitic" basalt. A magma that begins to crystallize deep underground but then erupts suddenly at the surface will have constellations of large crystals afloat in a fine matrix—the rock texture called "porphyritic." This is the story encrypted within my calligraphic rocks.

Unlike flamingos or mangroves, most rocks have life expectancies of tens to hundreds of millions of years, during which they will find themselves in a variety of different environments. Contrary to their reputation as insensate and impassive, rocks are attuned to their surroundings, and they alter their appearance accordingly, most often with the assistance of water.

Water is a resourceful makeover artist, both at Earth's surface and below. It picks up gases from the atmosphere and pilfers elements from rocks it encounters underground, then uses these to tint other rocks with new hues—like rusty red or forest green—depending on ambient chemical conditions. If groundwater has empty space to work with, it can transition from painting to sculpting, constructing three-dimensional mineral masterworks. Vugs left by gas bubbles in vesicular lavas provide the ideal studios for such creations, resulting in colorful "amygdaloidal" basalts dotted with minerals that are genetically distinct from those in the host rock. This is the origin of the orange starbursts that caught my eye (zeolite, I would later learn—a mineral used in water softeners) and also of the celebrated Lake Superior agates.

Depending on characteristics acquired during its genesis and subsequent history, each rock will also respond differently late in life to adversities like, say, being dragged from northern Ontario to Minnesota at the base of an ice sheet or pummeled by Lake Superior storm waves for ten thousand years. So it's not surprising that as a second-year student, still learning the basic grammar of rocks, I failed to recognize basalt in its many idiosyncratic varieties.

Rock identification is only one aspect of becoming geologically fluent. There are other, more challenging cognitive practices—indeed, deliberate retraining of the human mind—necessary for thinking like a planet. Among these are developing a radical reconception of time, and recognition that the boundary between past and present is highly permeable. A corollary is the habit of seeing the landscape not as a static fact but as an evolving palimpsest text that has been overwritten many times, in a never-ending process of revision. Accepting, in other words, that all geographies are temporary. A single region of Minnesota had been the stage for an empire of volcanoes, a vast inland sea, mile-thick lobes of glacial ice, and, at least for now, the greatest Great Lake. Building such understanding initially from the vantage point of a particular place allows one to apply it later to any spot on the planet.

At the time that I was learning to think this way, my personal sense of place was also changing. As a country mouse living in the city for the first time, I savored some aspects of urban life; opportunities for my continuing study of the human race were greatly expanded, and I enjoyed the anonymity of metropolitan spaces, where I could blend invisibly into the background. I made a systematic project of exploring various neighborhoods in Minneapolis and Saint Paul on my bicycle, visiting as many public libraries as I could and testing out cozy nooks. My favorites were the Carnegie buildings, monuments to a time when even ruthless industrial magnates valued literacy and gracious civic architecture. But welcoming as these were, they could not provide the same warm embrace as the sandstone walls of Mabel Tainter Memorial in my hometown.

As autumn advanced, I realized that my expeditions would soon be curtailed by snow and darkness. Back at home, I always welcomed the first snows of winter, eager to strap on my cross-country skis and glide from our front door into the brilliant white world. Now, the prospect of a claustrophobic winter in a stuffy dormitory was too much to bear, and with uncharacteristic boldness, I decided to see if the university's Nordic ski team might take me on.

Although I had never skied competitively, they did.

Most of the other team members had skied for suburban high schools in the Twin Cities area and were well-versed about the latest gear and training regimens. I still had wooden skis, woolen clothes, and had not even seen groomed trails more than a few times. Despite my subpar equipment and self-taught technique, I was embraced by the group.

Our coach was a wiry, elfin Norwegian American man who had competed in the 1952 winter Olympics in Oslo. In the weeks before the first snowfall, he had us build our stamina by "hill-bounding"—running up the steep banks of the Mississippi River near campus with ski poles. His coaching guidance was terse and almost philosophical; he explained that letting the body rest on the way back down the hill was as important as the exertion of going up. And the same applied to each stride on skis: one leg and arm should be relaxed while their opposites propel the body forward. In retrospect, I realize that this was good advice for living in general: learning to find interstitial moments of rest and calm in the midst of everyday struggles.

We were hill-bounding one warm autumn afternoon, and I was red-faced and sweaty in that puffy way that seems to come with Scandinavian genes. A striking, dark-haired team-mate who didn't seem to be suffering from the heat offered me a drink from his water bottle. The next practice was to be at a site farther from campus, and when I said that I didn't have a car, he replied that he was already taking a few of the others and could pick me up. Sometime after that, he mentioned that there was to be a winter sports equipment show at the expo center downtown. Maybe I could find a better pair of skis at a reasonable price. He could meet me there to help find a good deal. It took me several weeks of similar invitations to realize that he was asking me out.

His good looks came from his Greek roots, and I soon found myself amid a tight-knit Eastern Orthodox community that was foreign and fascinating to me. Siblings, cousins, aunts, uncles, grandparents, and godparents formed an enveloping mesh. There were women named Aphrodite and Athena—as if it were quite normal to be a goddess in human form. I was intimidated by, and vaguely envious of, the regal way they wore their womanliness, especially the mothers of multiple sons, who conveyed a particular type of matriarchal authority. "You're going to be a geologist? How interesting," they'd say to me politely. I can only imagine what they thought of my wardrobe, which consisted then of a collection of bib overalls in a range of fabrics and colors. My really fancy ones were green corduroy.

I learned to make spanakopita and avgolemono soup and was embarrassed to think of the prideful way Norwegian Americans urged lutefisk—slippery, translucent, lye-soaked cod, intended only to ward off starvation—on outsiders.

Stuffed grape leaves and baklava were truly food for body and soul. Services in the golden-domed Greek Orthodox cathedral, with mesmerizing icons, hypnotic chanting, and heady incense filling the sanctuary, moved me in a visceral, sensory way that the pallid Lutheranism of my youth never had. I felt myself drawn away from my family as I became more deeply immersed into this ancient, mysterious culture whose members were certain of their roles and destinies.

At home, my sister was also pulling away. For her, the normal adolescent struggles to define one's identity were magnified by her adoption. She directed her anger at our parents and the world at large and got entangled in the juvenile court system. I could understand the depths of her fury, the inherent asymmetry in our experiences of the world.

I had sometimes felt that our family, in hoping to provide not only a stable home for my sister but also help her understand her Ojibwe heritage, had taken on the burden of centuries of colonial history. Now, no longer able to hold up under that impossible load, our family felt close to splintering. My sister and I had lost our easy childhood rapport and seemed headed in opposite directions, while our parents withdrew into themselves to avoid further hurt.

Minneapolis got a lot of snow that year, and ski races provided good excuses not to go home on weekends. The team did well in both collegiate and citizen competitions, and I accumulated a shelf full of kitschy trophies. Then the season ended, and we put our skis away. On one of those soft spring nights when anything seems possible, my boyfriend and another member of the ski team struck on the idea that bicycling around Lake Superior would be good off-season training. How grand to be able to say we circumnavigated the largest lake in the world! Soon after classes ended in mid-June, we set off on our ten-speeds with skinny tires, carrying tents, sleeping bags, a camp stove, a change or two of clothes, and bags of oats, rice, and lentils.

The distance around the lake by road is more than eleven hundred miles. About halfway around, east of Nipigon, Ontario, at the pointed northern tip of the lake, we are at a low ebb. A cold rain is falling. Here the only road is the Trans-Canada Highway, and we've lost count of how many bone-rattling logging trucks have barreled past us. We haven't been sleeping well without any padding between us and the stony ground. We vastly underestimated our caloric needs, have almost no food that can be eaten raw, and only about fifty dollars in cash among us to last the rest of the trip. In any case, the next place that may have even a gas station is at least thirty miles away. And now we've just hit another long stretch of construction where the roadbed is armored with monster gravel—fist-size chunks of sharp rock that shred bike tires. I recognize it as basalt, and at this moment I loathe it.

There is nothing to do but plow on, each of us in the furrow of our own thoughts. The word "bedraggled" floats into my consciousness as an apt term for our sorry state. That in turn evokes a memory of when my sister first encountered the word in a book and read it as "bed-raggled"—a brilliant description of insomniac exhaustion. I smile in spite of myself. Eventually the pavement resumes, and the sun looks like it might come out. We grind our way up the next hill, and as miserable as we are, the beauty of Superior stuns us once again. Suddenly we can see ourselves in miniature—bedraggled, bed-raggled little creatures creeping around the margin of this gigantic, gorgeous lake—and collapse into giddy laughter at the juxta-position of the wretched and the sublime.

Having measured Lake Superior in this bodily, visceral way—its perimeter comes to just under thirteen days of aching legs—we cannot imagine that it hasn't been here forever. The vastness of the lake suggests permanence. But its apparent timelessness is an illusion; this very Great Lake is protean, never settled in a final form. It has no absolute moment of origin, only an evolutionary series of incarnations.

The water itself in Lake Superior is replaced about every 192 years. This is the average "residence" time for a drop of water in the lake, reflecting the relatively small rates of inflow and outflow compared with its huge volume. In other words, much of the water currently in Lake Superior fell as rain long before iron ore processors dumped tons of asbestos mineral waste into Silver Bay, Minnesota, in the 1950s and before the ill-advised Nipigon diversion in Ontario in the 1940s. Some of the water drifted down as snow on old-growth white pine forests of northern Wisconsin before the clear-cutting of the 1880s and 1890s, and some water still in the lake today flowed in via tributary rivers even before Ojibwe tribes were forced to sign the 1837 treaty that displaced them from ancestral lands. Lake Superior has a long memory, but soon it will forget the times before sawmills, mines, and cities. For some reason, I find this unspeakably sad.

The modern outline of Lake Superior is of course older than the water it contains, platted several thousand years ago after glaciers had radically remodeled earlier river systems. The ice lobes, like giant bulldozers, razed ancient drainage divides and constructed dams—ridges of rock debris called moraines—that would later capture their own meltwater. Several older versions of the lake can be recognized from abandoned shorelines at various elevations, left like giant bathtub rings on the landscape. Climate records from deep-sea sediments suggest that Northern Hemisphere ice sheets waxed and waned indecisively at least thirty times for more than 2 million years, and we can surmise that there have been at least that many iterations of Superior and the other Great Lakes. But because each new ice advance largely erased the record of the previous one—like successive generations of urban planners with different visions for a cityscape—only the most recent geographies can be discerned.

The topographic basin occupied today by Lake Superior is far more ancient than the Ice Age, however. The area that now holds 10 percent of the planet's fresh water is a long-lived low spot that has probably held many previous water bodies, sweet and salt, over the eons. It lies on the site of a billion-year-old tear in Earth's crust.

For decades, geologists knew from outcrops in Minnesota, Wisconsin, the Upper Peninsula of Michigan, and Ontario that the Lake Superior basalts formed an unusually thick sequence, a stack at least five miles deep. But even in the early 1980s, there wasn't a consensus about how this volcanic episode fit into the larger story of North America's construction, or within the still-new paradigm of plate tectonics. A map of gravity values in the central United States was the first clue that the same basaltic rocks exposed around the shores of the lake were part of a continental-scale tectonic event.

Most people think of the pull of gravity on Earth, 32 ft/second/second g, as a fixed value, but in fact it varies by tiny amounts over the course of the day because of the changing pull of the moon and sun, and by slightly larger amounts from place to place depending on elevation and rock type. Basalt is an especially heavy rock, rich in iron. Its specific gravity—its density relative to that of water—is around 3.0, compared, for example, with 2.7 for granite and 2.4 for typical sedimentary rocks. The basalts of the Lake Superior region are so dense and thick that their extra mass causes the pull of gravity to be a little greater than in surrounding areas. That is, you actually weigh a little more near the lake! The difference in gravity's tug is small—measured in thousandths of the average value of g—but enough to be detected with sensitive geophysical instruments (and to make our overloaded bicycles incrementally heavier on our trip around the lake).

A compilation of gravity measurements from the upper Great Lakes region to the Great Plains reveals a prominent stripe of elevated gravity stretching from the northeastern tip of Minnesota seven hundred miles southwestward into central Kansas,[2] where placid limestones at the surface convey no hint of volcanic unrest. At the time that I was picking up basalts on my first field trip, this swath of anomalous gravity was still being called the "Midcontinent Gravity High" (a great name for a school, I always thought—"Go Heavyweights!"). The uninterrupted character of the gravity high indicated that the same dense rocks that occur on the shores of Lake Superior must continue in the subsurface far into the heart of the continent.

But the truly immense volumes of basalt under Lake Superior were not appreciated until 1986, when a geophysical project called the Great Lakes International Multidisciplinary Program on Crustal Evolution or GLIMPCE[3] carried out shipboard seismic reflection surveys to image the crust below

the lake bottom. The resulting cross sections revealed that near
the center of the lake, basaltic lavas form a stack that is an incredible *fifteen miles* thick, three times greater than what can
be mapped along the shores. The seismic data also revealed
that the ancient preexisting crust in the region, the Canadian
Shield, with rocks normally about twenty miles thick, was
stretched and thinned in the area under the lake to less than
three miles. This was unambiguous evidence for the tectonic
context of the basalts: they had oozed up out of a continental
rift—a great gash in the crust—along which North America
had nearly been torn apart. The Midcontinent Gravity High,
a geophysical concept, became the Midcontinent Rift, a geologic province.

Together, the geological and geophysical clues reveal that
what is now the basin of frigid Lake Superior was a seeth-
ing sea of lava a billion years ago. Flow upon flow emerged
from fissures in the crust, cooling and hardening into basalt in its many varieties. Some of the flows are so extensive
and distinctive that they have names: the Osceola, Iroquois,
and Kearsarge. Ramparts formed by one batch of flows often dictated the paths of subsequent ones, and in some
cases created lava lakes as deep as the future Lake Superior. One such ponded flow, known as the Greenstone flow,
reached a thickness of 1,300 feet and today forms both the
spine of the Keweenaw Peninsula—the "thumb" of Michigan's Upper Peninsula that pokes into Lake Superior—and
the backbone of Isle Royale, which is the largest island in the
largest lake in the world.

Some flows bore up crystals from subsurface magma
chambers, forming porphyries with their cryptic calligraphy.
Frothy flow tops became spongy vesicular basalts that ground-

water would later transform into amygdaloids with spectacular agates and starburst zeolites. In the Keweenaw, heated groundwater bearing a rare cargo of dissolved metals seeped through the rocks and left rich deposits of "native" copper and silver—i.e., in pure elemental form—in the porous basalts.

If the crust had continued to stretch, breaching the old Canadian Shield rocks completely, a new ocean basin would have been created, the volcanic activity would have continued as a submarine seafloor spreading center, and the geography of North America would have been utterly different. But one geologic day, the magma chambers were depleted and volcanism ceased. The great accumulation of dense basalt caused the crust to founder like an overladen ship. Ever since, this crustal sag has shaped the evolution of the region, dictating the forms of Paleozoic seas, the paths of Pleistocene glaciers—and, eventually, the contours of the region's human history, which has always been bound up with basalt.

On our long bike ride around the lake, amid the vast wild tracts of forest, the immensity of the geologic past dwarfed our tenuous present. We and the thin ribbon of road were transient interlopers. How could we imagine that our "now" was more important, simply by virtue of being the most recent, than the eons of nows that had come before?

For thousands of years before the arrival of Europeans in North America, Indigenous people used copper from Lake Superior basalts for spear points, fishhooks, harpoons, and awls as well as bracelets and rings. Materials from archaeological sites where the earliest copper artifacts have been found date back to more than 7,600 years, meaning that the rocks of the Midcontinent Rift fostered one of the first known metalworking cultures in the world, at least a millennium before Lake Superior assumed its current form.[4] For many centuries after that, Lake Superior copper—identifiable by its unique mix of other trace metals, the signature of ancient groundwater—was traded via waterways and overland routes among Indigenous peoples across North America, perhaps the first commodity to be exchanged on a continental scale.

With about 150 miles to go on our circumnavigation of the lake, we cross the Ontonagon River, whose name is a synecdoche for the whole story of white seizure of Indigenous lands. When Europeans reached the area in the 1600s, they became feverish with copper lust. One particular chunk of Lake Superior copper, called the Ontonagon Boulder, is the embodiment of that sordid saga.[5] The boulder, a massive 3,700 pounds of nearly pure copper, had probably been transported from the Keweenaw to the south shore of Lake Superior by glacial ice. For twelve thousand years, the rock had sat on the banks of the Ontonagon River, about twenty miles upstream from where it flows into Lake Superior. Over time, the copper boulder had become a landmark and sacred object for the Ojibwe people in the region, who showed it to a Jesuit missionary in 1667. A British fur trader came upon it in 1766, made note of its impressive size and purity, and managed to scrape off a lump. Around 1820, Ojibwe guides led Henry Rowe Schoolcraft, an agent of the US government, to the boulder. Schoolcraft, a self-taught geologist and savvy speculator, suspected that it had come from rich in situ deposits of copper not far away.[6]

In 1826, based partly on Schoolcraft's reports, the US government compelled the Great Lakes tribes to sign the Treaty of Fond du Lac, ceding all mineral rights in what is now Michigan, Wisconsin, and Minnesota. By 1837, the boundaries of the new state of Michigan had been drawn in an elaborately gerrymandered way to include not only the Upper Peninsula (and especially the copper-rich Keweenaw) but even Isle Royale on the far western side of Lake Superior, which had also been found to host copper-bearing basalts.

Meanwhile, sometime in the early 1830s, a Detroit shopkeeper, one Julius Eldred, believed that he had rightfully bought the Ontonagon Boulder from the Ojibwe and tried unsuccessfully for years to move it. In 1843, the indefatigable Eldred returned to the Ontonagon River with a hardy crew and more sophisticated equipment, only to discover that the US secretary of war had sold the boulder to a hard-bitten group of miners from the Wisconsin Territory. Copper is malleable but not brittle; it is almost impossible to knock off a sample with a hammer, and the miners had likely given up trying to break or move the boulder. They were probably happy when the persistent merchant offered to buy the boulder he considered his own back from them for ten times what he had paid to the Ojibwe. Eldred and his gung-ho gang of twenty hoisted the boulder to the top of the riverbank and built a temporary rail line to bypass unnavigable rapids on the Ontonagon, laying down and tearing up the same short stretch of track in a leapfrog pattern until they reached a spot where a boat could carry the boulder out to Lake Superior.

Some of the water that is still in the lake today witnessed these events.

Back in Detroit, Eldred put the boulder on display, charging admission for a glimpse at his prize, and raising public awareness of the vast riches lying within the rocks of the wild Lake Superior shores. The US War Department then determined that the merchant's (re)purchase of the boulder had not been legitimate and took the matter to court. The government ultimately prevailed but paid Eldred about three times the amount he had paid the miners for the boulder, as compensation for the costs of moving it. Several years later, the Ontonagon Boulder was turned over to the Smithsonian Institution, where it remains in storage today, even though the Keweenaw Bay Ojibwe Community has been petitioning since 1991 to have the boulder returned to the place where the glacier left it.

Around the time Eldred was wrangling the Ontonagon Boulder onto a boat, the War Department—whose involvement is a recurring theme in this story—became concerned about protecting the copper country from various unauthorized plunderers, including Native Americans. In 1844, the army established a fort at Copper Harbor on the remote tip of the Keweenaw Peninsula and named it for the serving secretary of war. The soldiers based at Fort Wilkins faced no uprisings or insurrections in the subsequent decades, but the Keweenaw Peninsula itself was subjected to violent assault, in North America's first metallic mining rush.[7]

Between 1845 and 1938, the Keweenaw was one of the most intensively mined areas in the world. More than 250 distinct mining operations tunneled into the many rich copper lodes—the foamy vesicular tops of ancient lava flows—that tilt northwestward toward the lake. Fortunes were made and lost. The promise of good-paying jobs drew people from all over the Old World to this snowbound peninsula, and in the 1910s, newspapers in Finnish, French, Italian, and Slovenian were published daily in the city of Houghton.[8] Mine disasters and shantytown fires forever altered family histories. Labor unions rose and were crushed by company bosses. In the great strike of 1913 to 1914, the legendary labor crusader Mother Jones herself marched with the copper miners. Behind all this toil and tumult, all the claim-staking and hardscrabble living, was basalt. The story of the Keweenaw challenges the notion that rocks are inert matter over which we have control. Rocks have the power to inflame our minds, incite violence, and alter the trajectory of history.

By 1920, the most accessible copper deposits had been mined out. The Keweenaw was a gutted, inside-out landscape—the surface a wasteland of broken rock coughed out of the mines and the subsurface a warren of collapsing shafts and rooms. The mighty Quincy Number 2 mine, "Old Reliable," was the most consistently productive of them all. It had reached a depth of seven thousand feet, more than a mile below the water table, and so deep that its skip system could no longer manage to haul waste rock and copper up to the surface. The company commissioned a specially designed steam hoist, the largest in the world, from a company in Milwaukee and constructed a gleaming building to house it, a glorious temple to capitalism, with graceful lunette windows and interior walls of white Italian tile. At the time that the giant new hoist started operating in 1920, the Quincy company had paid dividends to stockholders every year since 1867. The company distributed its last dividend in 1921.[9] By 1931, it was bankrupt, defeated in the end by competing mines in Montana that did not have to contend with such great thicknesses of basalt.

So where did all of this basalt, whose irresistible force field has warped history, come from?

Although rocks don't map neatly onto a simple evolutionary tree, basalt can be considered more "primary"—closer to the trunk—than most others. In fact, basalt is the most common rock type not only at the surface of the Earth—it lines the floor of the world's oceans—but also on our neighboring planets and in the dark "seas" on the moon. When students in my classes express frustration at the profusion of rock types that they are expected to know, I joke that maybe they'd be happier joining certain tech titans on Mars or the moon, where there are only a few rock types, including basalt. There, class could be wrapped up in a couple of weeks.

In my own student days, I often felt a dissonance between the flat affect with which our professors delivered their lectures and the profound implications of what they were saying. This planet, I realized, is truly strange. We Earthlings take for granted the great variety of rock types that create the geologic infrastructure in which we live. But this is only because Earth is such a versatile and virtuosic cook, a planet that has learned to mix, distill, and recombine available ingredients in ways that no others have. Basalt, however, is one of the most basic recipes, whipped up in two easy steps in hell's kitchen: just heat up some mantle rock, then take the lid off. Even amateurs like the moon, Mars, and Venus have mastered that. There are a few more details to the technique, but the recipe is simple. The first step is to gather the raw materials.

Our solar system formed 4.55 billion years ago from elements forged by ancient stars, and all the rocky planets and moons in our solar system started with the same pantry of ingredients. First, our sun scavenged enough recycled matter to flare into being, and then in the high-temperature region closest to the new star, comparatively scarce heavy elements—iron, magnesium, silicon, aluminum, calcium, and many others—condensed into granules. These planetary seeds accreted into larger bodies, which melted and then separated by density into iron-nickel cores and silica-rich, rocky mantles with lower concentrations of these metals. Mercury, Venus, Mars, Earth, and the moon all organized themselves this way.

Such segregation into core and mantle is permanent; no superheavy core material could ever rise through the less-dense mantle and reach the surface. But a planet's core may still communicate with the mantle by sending heat through the boundary that divides them, like propaganda beamed across a closed political border. This heat may in turn cause instability, even in a fully crystalline rocky mantle made of silicate minerals. Silicates expand significantly when heated and thus, like the wax in a lava lamp, will rise buoyantly, flowing in the solid state toward the surface at the impressive pace of an inch per year. This is called thermal convection, and setting up such a convective system is the first step in making a batch of basaltic lava.

The second step is counterintuitive but incredibly simple: just reduce the pressure. This can be done by thinning the crust, as in the Midcontinent Rift (Earth's preferred method) or by having large meteorites punch holes in it (lunar method, not recommended). *Et voilà!* Gooey liquid basalt will ooze out at the surface. The recipe is time-tested and fail-safe. Even if your mantle rock is cooling it will still be plenty hot to melt. It seems like magic, but the explanation is simple: the melting temperature of silicate rocks (like all normal solids) decreases with decreasing pressure (water ice, though more familiar, is anomalous and the relationship is opposite). So, if you have rock that is close to its melting point and decompress it, it will melt. You might, however, note that your mantle rock has not completely melted and that residual crystals remain.

This is normal and essential to the process; these are the green mineral olivine (or peridot), which melts only at very high temperatures. The mantle consists of an array of components with different melting points. Decompression melting tends to extract those elements that melt out at the lowest temperatures, and upon cooling, these will reliably crystallize to plagioclase and pyroxene, the key mineral ingredients in basalt.

This process of decompression melting created the smooth basaltic lava plains of the moon's maria basins, including the Sea of Tranquility, where Neil Armstrong stepped off the lunar module. Basalt was thus the first rock humans encountered on another world. Venus too, though shrouded in a dense fog of volcanic carbon dioxide, seems to be lacquered with basalt from pole to pole. Its comparatively uncratered surface suggests that these lavas are fairly young and that decompression melting may still be happening just beneath its cloud-veiled surface.

Mars, likewise, seems to be a world of basalt. Rovers and orbiters have probed and mapped it and found nothing but basalt, basaltic sediments, and basalt altered in places by long-vanished groundwater. About 250 rare meteorites that are thought to have come from Mars (based on trace concentrations of noble gases matching those in the Martian atmosphere) are mainly basaltic. And Mars's gargantuan extinct volcano, Olympus Mons, the largest in the solar system, has the telltale shape of a basaltic shield volcano like Hawaii's Mauna Loa, with gently sloping sides that reflect the runny, low-viscosity nature of basaltic lavas.

Like the Hawaiian volcanoes, Olympus Mons seems to have been built by basaltic lavas formed when an upwelling plume of mantle rock underwent decompression melting. But at sixteen miles high, it is three times taller than the height of Mauna Loa measured from the sea floor. The gargantuan size of Olympus Mons reflects a fundamental difference between Mars and Earth: The crust of Mars is a thick, cold rind of basalt, a single global plate that doesn't move relative to the underlying mantle—and apparently didn't move even when the mantle was still hot enough to convect. So, on Mars, a rising column of mantle rock undergoing decompression melting would just keep extruding basaltic lavas at the same spot on the surface, constructing titanic volcanoes like Olympus Mons.

Earth, in contrast, has a relatively thin crust broken into plates that move relative to one another and to the mantle below, and basaltic lavas formed by decompression melting don't usually have time to accumulate into such immense piles before the overlying plate leaves the scene. And, in fact, most Earth basalts eventually return to the mantle via the crustal recycling process of subduction—the downwelling part of the thermal convection system. Earth alone practices "active-lid" convection, in which not only the mantle but also the basaltic ocean crust is involved in the slow thermal overturn of convection. A planet may have basalt without plate tectonics, but no plate tectonics without basalt.

Yet the thickness of the Lake Superior basalts—fifteen miles, as shown by the GLIMPCE seismic surveys—rivals the height of Mars's Olympus Mons. This reflects another way in which the Midcontinent Rift basalts are exceptional: the extraordinary *rate* at which they were extruded. When high-precision isotopic dates for Lake Superior's basalts were determined for the first time in the 1990s, they revealed that the bulk of the lava was spewed out in a geologically brief interval of just a few million years between about 1,102 and 1,098 million years ago (as it happens, this was about the time that the last major eruptions were occurring on Olympus Mons). Such an effusive rate of eruption qualifies the Midcontinent Rift as one of the great "flood basalt" events in Earth's history.

Flood basalt episodes—feverish intervals in which more than a quarter million cubic miles of lava are erupted onto land in a limited area in a few million years—are, fortunately, rare, because they seem to be very bad for life on Earth. The greatest mass extinction in Earth's history, the end-Permian cataclysm 250 million years ago, seems to have been set in motion by the frenzied eruption of flood basalts known as the Siberian Traps. The end-Triassic extinction, which ranks as the fourth worst, was similarly linked with an out-of-control period of basaltic volcanism at the time that the modern Atlantic Ocean began to form through continental rifting about 200 million years ago. And even the end-Cretaceous demise of the dinosaurs, which is famously pinned on a rogue asteroid, coincided with the eruption of flood basalts in India, known as the Deccan Traps.

The Earth is constantly churning out basalt through decompression melting along the global chain of submarine volcanoes called the midocean ridges, in the process of seafloor spreading. This process is fundamental to the planet's plate tectonic system. But flood basalt eruptions are anomalies far beyond Earth's business-as-usual basalt output. Neither their cause nor the reasons for their especially negative effects on the biosphere are well understood. Most geologists concur that it was probably not the lavas themselves but the associated volcanic gases—especially excess carbon dioxide—that most likely destabilized the climate and ecosystems and led to mass extinctions of the past.

Could the eruption of flood basalts from the Midcontinent Rift have caused a mass extinction? The vesicular flow tops that bore copper, made fortunes, and broke miners' backs attest to the great volumes of magmatic gases that must have emanated from the roiling sea of lava in the Rift. But the Lake Superior basalts erupted on an Earth whose biosphere consisted only of single-celled organisms, and detecting a mass extinction in the fossil record of microbial life is difficult. And even an event as extreme as the Midcontinent Rift volcanism may not have flustered a simpler, more adaptable biosphere that did not yet have elaborate food chains—or precarious economies based on the delusion that Nature is inert and unchanging.

In a lovely geochemical irony, flood basalts whose excessive carbon dioxide exhalations once wreaked havoc with global climate may now help draw down human-emitted carbon dioxide levels that threaten to destabilize civilization. Basalt, native to Earth's mantle, is far out of equilibrium with conditions at the surface of the planet. In the presence of water made slightly acidic by carbon dioxide, the primary minerals in basalt, plagioclase and pyroxene, release calcium, which can recombine with the carbon dioxide to form calcite, thereby fixing the atmospheric gas in solid form. Recent experiments indicate that amending agricultural soils with pulverized basalt could be an effective carbon sequestration strategy that also boosts crop yields.[10] The vast waste-rock piles from the Keweenaw could be "remined" in a way that actually helps to reduce the human imprint on the planet.

It's rather extraordinary that any basalt formed so long ago—at a time when the mix of gases in Earth's atmosphere was entirely different—could have a vigorous conversation with the air today. Most basalts don't live long enough to see such changes. Seafloor basalts, whose destiny is to be recycled into the mantle by subduction, have a residence time at the Earth's surface of about 150 million years. This might have been the fate of the Lake Superior basalts if the Midcontinent Rift had stretched the crust just a bit more and opened a new ocean basin. Instead, the old Canadian Shield did not break apart, and Lake Superior's basalts, safely aboard an unsubductable continent, have survived long enough to witness the global basaltic ocean floor replace itself five times. If the Rift had ruptured North America, the subsequent choreography of the plates, evolution of land-based organisms, and eventual geographies of human history would have been entirely different.

In the summer before my senior year in college, I went away to "field camp," once an essential rite of passage for all geology students. At that time, the University of Minnesota held its camp high in the Sawatch Range of Colorado, and for six weeks, we were an insular little community living in tents at eight thousand feet, spending every daylight hour mapping and interpreting rocks, expanding our geologic vocabularies well beyond basalt. My cross-country ski training served me well and earned me some respect among male classmates who were always competing to be the first to reach summits. I also became a champion at the unofficial sport of "boot-skiing," a secret thrill practiced by many generations of field geologists. One may spend all day climbing up a mountainside, but if there is a slope mantled with small pieces of broken rock, it can be descended in long, gliding strides in a few exhilarating minutes.

On Saturdays, we would drive an hour into the nearest town to take showers, do laundry, stock up on beer, and enjoy other amenities of civilization. One Saturday night, a group of us went to see the new movie *Blade Runner* in the town's once-grand theater on the wide main street. I recall feeling distressed, almost poisoned, by the film's depiction of a dystopian urban techno-future. Stepping out of the theater into a Rocky Mountain night infused with the incense of pine, I was filled with joy and relief at being alive in that place and time on a verdant, vibrant planet.

Field camp had reawakened the sense of wild freedom in nature that I had had in childhood, and when I returned to Minneapolis, my boyfriend's clannish Greek community seemed claustrophobic. I had longed to be enveloped by a collective, to relax into the certainty of a life that was already platted out. But I realized now that such an embrace would grow to feel stifling, and that within that community, I would never fully assimilate nor ever unfold into myself. Although the mysticism of the Orthodox church still drew me, I now sensed the overtones of misogyny that came with it. I was beginning to realize that for me, the deepest truths were to be found in nature, not in human traditions. The asymmetry between the well-established edifice of my boyfriend's beliefs and the embryonic state of my own meant that my worldview would always be eclipsed by his. Feeling shackled and constrained by his vision for our future, I was irritable with him, and we broke up acrimoniously. I had stretched as far as I could toward an alternative destiny, but my own center of gravity pulled me back toward a rockier path.

Long after my first field trip to the shores of Lake Superior, when I was just beginning to learn the language of rocks, the lake's gravitas still draws me in, its basalt a touchstone in uncertain times. I realize now that this greatest of lakes is an avatar for the planet as a whole, ancient but ceaselessly renewed; perpetual yet also provisional. And its rocks exert a force field that has shaped the arc of history in ways that humans, preferring to believe we control our own destiny, barely grasp.

Basalt, the most common rock type in the solar system, from the impact basins of the moon to the barren plains of Mars to the floor of the oceans and the shores of Lake Superior, is a solid foundation for a planet and a good primer for a geologist. But on Earth, basalt is only the starting point.

3

||||||||

TUFF

It's 1:00 a.m. on a starry July night, and I'm driving a government pickup with a bad muffler along a gravel road on the east flank of the Sierra Nevada mountains in California. I hadn't expected to be sent out by myself so early in this internship with the US Geological Survey, but I'm determined to show my male supervisor that I am competent and tough. Only a few weeks have passed since my college graduation in Minnesota, and everything here still feels alien—the air carries a scent I can't identify, the plants bear names I don't know, and the landscape has a raw, unfinished look that eludes my interpretation. As a native of the stable continental interior, I'm not used to places where tectonic forces are still actively shaping the terrain.

Working in the middle of the night is even more disorienting. I'm not really sure if I'm heading the right way. I peer into the darkness, trying to recognize any distinctive trees or rocks from the previous day, when we installed a brass survey marker at a site with a clear view out over Long Valley and beyond, into western Nevada. There are more intersections than I remember seeing in the daylight, and I've lost track of how many turns I've taken. It's 1983, pre-GPS, and all I have is my sense of direction and a highway map that doesn't even show these back roads.

A rising sense of panic competes with my stubborn pride. There is a walkie-talkie, but calling my boss to say that I'm lost on this very first outing would be an admission of failure. My mind races ahead to worst-case scenarios: I mess up tonight, he writes me off as a ditzy woman, and my career is over before it's even started. I'm blinking back tears when I spot some blaze-orange tape we tied on a tree to mark the turnoff. Relief surges over me. My supervisor's voice rasps on the radio: "You there yet?"

He and a technician are setting up a laser instrument more than ten miles away, on the opposite side of Long Valley. The valley is actually a caldera—the gutted remains of a supervolcano that blew itself up 760,000 years ago in a cataclysmic explosion of a magnitude similar to the more famous ancient eruptions at Yellowstone. In recent months, a swarm of earthquakes (to use the preferred collective noun) within the old caldera has caused geophysicists to speculate that magma could be rising through the crust.

Similar earthquake clusters occurred just before the eruption of Mount Saint Helens in Washington State three years earlier, and the apocalyptic scenes from that event are still vivid in everyone's minds. Our task is to make high-precision measurements of the distance across Long Valley to determine if the ground is bulging upward, which could portend an impending eruption. To achieve a resolution of inches across the width of the valley, we need to work at night when the atmosphere is less stratified and the laser light will remain better focused.

I climb out of the truck and start unloading the equipment, hoping that no one else is on these roads at this hour. The possibility of encountering some rough-living hermit out here lurks in the back of my mind. I locate the survey marker set in a dab of concrete, one of a dozen we've installed in the past several days when we were still working in beautiful sunshine. We named the first few sites after notorious volcanoes, stamping the names letter by letter—*V-E-S-U-V-I-U-S*—in graceful arcs on the metal discs, until the lead geophysicist on the project said that we should use a generic numerical system in case someone from the already jittery public would come across these benchmarks. But this site remains: Krakatoa.

My next challenge is to set up a laser reflector mounted on a telescoping twelve-foot pole. I lay the extended pole on the ground with its base on the survey marker, then using ropes attached to the top and my foot to stabilize the bottom, manage to pull it into a precarious vertical position. Now I need to anchor the ropes with stakes, but there is almost no soil. The ground is made of rubbly tuff—rock made of volcanic ash and pumice fragments. I'll need to get a rock hammer from the truck to pound the stakes in, and to do that, I have to lower the pole back to the ground. Another crackle from the walkie-talkie: "You ready?"

Thinking now of my twenty-year-old self alone on a mountainside in the middle of the night, I feel some retroactive anger toward my boss: Was this hazing or just bad judgment on his part? At the time, I saw my fear as a symptom of my own weakness—a sign that I was probably destined to crash and burn and would end up confirming everyone's belief that a young woman couldn't handle such a job.

In the coming weeks, however, I become adept at hoisting the reflector quickly into position, and the tangled skein of back roads above the valley grows familiar. I learn the names of trees and shrubs and come to love the pineapple-scented bark of the Jeffrey pines. There is occasionally time during the day to explore the surrounding area, including Mono Lake, with its otherworldly towers of "tufa"—spongy calcium carbonate precipitated at sites where springs bubble up into the briny lake water.

Both tufa and tuff are new rocks to me, and also far newer—younger—than any I'd encountered before. Coming from the Lake Superior region, where the billion-year-old basalts of the Midcontinent Rift are among the youngest, it is strange to be in the company of rocks less than a *million* years old, still sitting in the environments in which they formed. "Tufa" and "tuff" share an etymologic root in the Italian word *tufo*, meaning "a soft, porous rock," but they could hardly be more dissimilar in their geologic origins—one is formed by silent seeps, the other in volcanic violence. To me, both seem immature, fragile, not really even proper rocks yet—but then I chastise myself for being a rock snob.

Working at night gets a little less strange and even affords some fringe benefits: One survey station is next to a popular hot spring where I can soak for twenty minutes in complete solitude once I get the reflector set up. And our lodgings are almost absurdly posh. With the Geological Survey's announcements about the possibility of a volcanic eruption, scores of condominium owners in Mammoth Lakes, a swanky ski resort town, have sold off their properties. Our per diem allowances are more than enough to cover rent for a luxurious cabin—with a hot tub, stone fireplace, and wraparound deck—on the flank of Mammoth Mountain.

The community itself is less accommodating; we get a chilly reception from year-round residents and business owners, who blame the Survey for causing the local economy to collapse. During the day, while doing errands in town, I avoid driving the USGS truck with the agency logo emblazoned on its doors, preferring to carry groceries on my back rather than risk confrontations in the parking lot with people who begrudge geologists. It feels unfair to be made pariahs simply for listening to Earth and translating what we are learning—though this is only the first of many times I will encounter people who resent geologists for talking about things they would rather ignore, like climate change, or groundwater contamination, or evolution.

Evidence of past cataclysm is so obvious to our eyes. The town sits on the west rim of the caldera, and the geothermal pools that draw people in all seasons are direct evidence of lingering magmatic heat. Pieces of pumice—glassy volcanic foam—float, improbably, in nearby lakes. Just to the south along the Owens River, the highway cuts into ash deposits more than six hundred feet thick. This massive ash layer is called the Bishop Tuff, after the nearby town, which would have been a Pompeii had the eruption happened in more recent times. But the Long Valley Caldera exploded long before the first people arrived in North America, and we can imagine its fearsome effects without having to contemplate horrors it caused for humans.

In the area around Long Valley, the Bishop Tuff records the deadliest of all volcanic phenomena: a pyroclastic flow. Also called a *nuée ardente*, or "burning cloud," a pyroclastic flow is a turbulent mix of incandescently hot ash, crystals, pumice, and searing gases, surging over a cushion of air at speeds that can exceed two hundred miles per hour. During the Long Valley eruption, such hellish flows coursed away from the caldera to distances of more than forty miles, leaving a distinctive stratum of "welded" tuff called an ignimbrite (a geologic neologism, Latin for "fire cloud rock"). Equally impressive is the spatial extent of the air-fall component of the Bishop Tuff, which is found as far east as the central Great Plains.

Geologists who have spent careers studying the Bishop Tuff estimate that a total of 145 cubic miles of magma was erupted in a single sustained eruption. This is thirty times more than Krakatoa in 1883, 180 times more than Vesuvius in 79 CE, and 2,400 times more than Mount Saint Helens in 1980. Emptying such a vast magma chamber would have required an unimaginably violent outburst that probably lasted a week.[1] No eruption of such a magnitude has happened in human history.

Although the eruption of Mount Saint Helens in 1980 was a tiny hiccup compared with great volcanic events in geologic history, it has outsize importance in the field of volcanology because it was the first to be documented instrumentally from the earliest hints of unrest through the main explosion. Saint Helens's first symptoms—earthquake swarms and steam plumes—began around the Ides of March, and from that point the volcano was monitored like a patient in a hospital, its vital signs carefully recorded around the clock. High-resolution seismic networks tracked magma pathways, geodetic surveys measured a growing bulge on the mountain's north side, and new instruments for monitoring gas emissions revealed a steady change in the volcano's exhalations, from mainly water vapor and carbon dioxide at the start to progressively higher levels of sulfur dioxide as weeks passed.

By early May 1980, geologists had developed a feeling for the volcano's idiom and knew it was signaling that a major

eruption was imminent. Some aspects of the ultimate eruption on May 18 were not predicted, most tragically the near-
horizontal north-directed initial blast, which killed one of the
volcano's closest observers, Dave Johnston of the USGS. But
the Mount Saint Helens event showed that volcanoes tend to
announce their intentions prior to major eruptions and that
human lives could be saved by heeding their communiqués.
In a sense, Saint Helens allowed geologists to develop a kind
of dictionary of volcanic phrases. The ability to interpret these
premonitory utterances led to the successful evacuation of
more than sixty thousand people before the far larger eruption
of Mount Pinatubo in the Philippines in 1991.

At the time Mount Saint Helens erupted, the unifying
theory of plate tectonics was just over a decade old, and geologists had only recently been able to explain why some volcanoes undergo such violent paroxysms while others—like
those in Hawaii and in Iceland—exude lavas with little drama.
By the late 1960s, geophysicists had noted a connection between the locations of the world's most explosive volcanoes
and sites of anomalously deep earthquakes in Earth's upper
mantle, where rocks should ooze, not break. This correspondence was especially evident around the margins of the Pacific
Ocean, with its volcanic "Ring of Fire" looping from Central
America and the Andes, across to New Zealand, north to the
Philippines and Japan, over to the Aleutian Islands, and back
around to the Cascades, including Mount Saint Helens. In
all of these places, deep earthquakes indicated that slabs of
abnormally strong—and presumably cold—rock were sliding
from Earth's surface into the mantle. A 1970 paper introduced the word "subduction" to describe the process occurring
at these sites, where basaltic ocean crust, late in its life and

far from its origin at a midocean ridge, has become cold and dense enough to slip back into Earth's interior.[2]

Subduction zones not only produce the planet's deepest earthquakes but also—closer to the surface—the largest and deadliest ones. On average, over geologic timescales, subduction occurs at the stately rate of inches per year. Over shorter intervals, however, the pace is unsteady, quantized into increments, because the upper part of the descending slab meets frictional resistance from the overriding plate. This shallow part of the slab may get stuck or "locked" for centuries before it lurches abruptly downward, sometimes by as much as a hundred feet in a matter of seconds, causing "megathrust" earthquakes like the giant magnitude 9 events that unleashed devastating tsunamis in Sumatra in 2004 and in Japan in 2011.

But why would these sites where *cold* rock is sinking into the mantle generate fiery magmas that erupt so violently at the surface? The answer, counterintuitively, is water. When ocean crust is forged by decompression melting at a midocean ridge, seawater surges through the fresh basalt, creating deepsea geysers called "black smokers" and triggering reactions that hydrate the rocks. A hundred million years later, that chemically bound water is carried with the now cold ocean plate as it slides back into the mantle. Once a subducting slab passes through the sticky locked zone, it is slowly warmed by the mantle's ambient heat, and at a depth of about thirty miles, the slab starts to sweat out the seawater acquired at its birth. That water, in turn, acts as a "flux" for the wedge of mantle rock above the slab, lowering its melting point and generating magmas of distinctive—and destructive—compositions that feed volcanoes like those in the Ring of Fire.

Because water is implicated in their very genesis, magmas at subduction zones are full of dissolved gas. They also have a different composition than basaltic magmas generated by simple decompression melting in places like Hawaii, Iceland, midocean ridges, or the Midcontinent Rift. The presence of water selectively melts out certain constituents of mantle rock, extracting elements like potassium and phosphorous—present only in trace quantities in the mantle—and carries these toward the surface, concentrating them in the crust. These water-generated magmas are also lower in density and higher in silica (SiO_2) than typical basaltic melts, which are about 50 percent SiO_2 by weight. Andesite magmas, common in Japan, are about 60 percent SiO_2 while rhyolite magmas erupted at Mount Saint Helens and Long Valley, among other places, are 70 percent SiO_2. These differences in chemistry have profound physical consequences: as silica concentration increases, the viscosity, or stickiness, of magmas rises exponentially. This occurs because silicon-oxygen molecules start linking up into long polymer chains—embryonic crystals—that tangle and interact, dramatically increasing the stiffness of the magma.

Magma, when rich in both water and silica, is an especially dangerous concoction. Magmas are almost always less dense than the solid rock that surrounds them, so they rise buoyantly through fissures and vents in the crust. As magmas near the surface, the gases dissolved in them begin to bubble out, in the same way that bubbles form when a bottle of champagne is opened. The formation of bubbles—called vesiculation—sets in motion a positive feedback cycle that leads to eruptions (and sometimes jet-propelled corks). Bubbles, being filled with gas, reduce the bulk density of the magma, so it rises higher, which allows bubbles to grow larger, and so on, until the molten rock reaches the surface. In a low-viscosity magma, this process may simply lead to effusive eruptions like the ones that produced the great thickness of basalt beneath Lake Superior. But in stiff, high-viscosity magmas like those formed in subduction settings, bubbles may not be able to escape. Gas pressures can increase to the point where they exceed the weight of overlying rock, and then literally blow the roof off in an explosion of rock shards and glassy, fragmented magma—aka ash.

Under the microscope, volcanic ash reveals the violence of its origins. Many ash particles have the peculiar form of a triangle whose sides are collapsed inward. This spiky shape is what makes ash so dangerous to breathe; the needlelike points can lodge deep into lung tissue. But the shape is also a sobering record of the last moments before an eruption: those concave sides trace the outlines of rapidly expanding bubbles in the magma seconds before it blew itself up completely. In our daily encounters with the immense volume of ash in the Bishop Tuff, I began to get some sense for the titanic scale of the Long Valley eruption and was glad to contemplate it from the safe distance of a few hundred thousand years.

Our rented cabin in Mammoth Lakes was commodious but sparsely furnished, and the kitchen lacked some essentials. One day I set off on a walk up the mountain and spotted a rummage sale being held outside a garage-like unit at a storage facility on the edge of town. I wasn't embarrassed to buy used goods; my childhood home was furnished with items from farm auctions. Since many people stayed in Mammoth only during the winter, it seemed natural that they might stash extra stuff in such units over the summer months.

But this collection of belongings seemed different. It appeared to represent the unsorted contents of an entire household, including personal photos. Based on the clothes and selection of music on cassette tapes, the owner would have been in his late twenties. Eavesdropping on people around me, I learned that the sale was being held by friends of a young man, estranged from his family, who had recently died of AIDS. I couldn't bring myself to buy anything. Every object seemed infused with unspeakable tragedy.

Until this point, AIDS had been an abstraction to me, happening elsewhere to no one I knew. In subsequent years, it would come nearer, claiming dear old friends from my hometown. But that sale on the rim of an ancient volcano was the first time I glimpsed, in the shards of a life, the cataclysmic effects of the epidemic.

The early 1980s are now viewed through the gauzy haze of nostalgia as a simpler, sillier time of shoulder pads and big hair, and it's easy to dismiss the apocalyptic feelings that hung in the air. The unstoppable horror of AIDS was playing out against a backdrop of heightened Cold War tensions. The eruption of Mount Saint Helens seemed like a rehearsal for nuclear holocaust. In news footage of forests felled like matchsticks, rivers deranged by mudslides, everything under a pall of gray ash, it was hard not to see the nightmarish aftermath of an atomic bomb. All of this amplified my personal sense that everything was precarious. My sister's troubles continued, and I was wracked with doubt about being able to chart an independent way in the world. My parents were supportive but wondered why I had chosen such a rocky path. The cozy stability of our 1970s household seemed long gone.

In the 1980s, Californians learned of the inevitability of
the "Big One"—a future magnitude 8+ earthquake on the San
Andreas fault. Earthquake science was undergoing a revolution. Seismographic data had evolved from analog records in-
scribed by pens on rotating scrolls to digital signals that could
be processed to determine the magnitude and origin of an
earthquake within seconds of its occurrence. Historic earthquake data could also be compiled and analyzed to look for
recurring patterns, and there was optimism that with enough
information, earthquakes, like volcanic eruptions, might one
day be predictable. In retrospect, that idea does seem to come
from a simpler, more innocent time.

After we had completed our baseline round of measurements across the Long Valley caldera, my USGS team was assigned to help with a project along the San Andreas fault in
central California, near the "town" (population twenty, rounding
up) of Parkfield. This town was famous among geophysicists because of its remarkably regular history of earthquakes; from the
time of an 1857 quake when recordkeeping began, this segment of the San Andreas had slipped in magnitude 6 events
every twenty-two years or so. In addition to its celebrity status
among seismologists, Parkfield has connections to the classic
days of Hollywood. William Randolph Hearst—the real Citizen Kane—owned a sprawling ranch just west of Parkfield.
And a lonely intersection a few miles south of town was the
site where, in 1955, the actor James Dean crashed his car, im-
mortalizing himself at age twenty-four as an icon of brooding
existential unease.

When our group arrived in Parkfield after a long drive
over the Sierras, all the rooms at the only local motel were
already occupied by senior geophysicists or people on James

Dean pilgrimages, and we ended up staying in an abandoned farmhouse on the old Hearst ranch that the resident rodents graciously shared with us for a week.

At that time, it had been seventeen years since the last Parkfield earthquake, in 1966, and if the pattern held, the next earthquake would happen soon. Seismologists had begun to descend on the town because it was the perfect place to investigate the physical factors governing the "stick-slip"' behavior of faults—the rock properties and processes that regulate their alternation between frictional locking and seismic sliding. And it was hoped that such understanding could lead to identification of precursory signals that might allow real-time earthquake prediction.

Our job at Parkfield was to help install instruments along the fault to track its moods at close range, in the earliest days of what would eventually become a much more ambitious project called the San Andreas Fault Observatory. In addition to highly sensitive seismometers, these devices included a "dilatometer" placed in a drill hole to measure tiny volumetric changes in the rocks that could presage an earthquake. It was known from laboratory-based "rock squeezing" experiments that rocks under stress exhibit surprising behavior just before they fail. Although they are being compressed, they actually expand in the moment before they fail brittlely. This reflects the opening of scores of tiny "microcracks" that gap apart in the minute or so before a rock fractures. This dilatancy, in turn, can lead to transient changes in elevation of the water table, as groundwater is sucked down into new pore spaces, and also to fluctuations in groundwater chemistry as radon and other elements trapped in the rocks are suddenly released. There was optimism that these and other phenomena, if closely

monitored, would eventually make it possible to issue earthquake warnings and watches akin to those for tornadoes.

The unusually consistent history of similar-size earthquakes
at Parkfield also gave rise to the "characteristic earthquake"
concept: the idea that certain segments of an active fault like
the San Andreas tend to slip again and again by the same
amount with a recurrence interval that is controlled by long-
term frictional characteristics of the rocks along the segment.
Further support for this idea came from reanalysis of data
from the two Parkfield quakes (in 1934 and 1966) that had
been recorded instrumentally: Specifically, the waveforms for
these events—the squiggly lines on seismograms, the "voice-
prints" of the earthquakes—were nearly identical, suggesting
that the fault had ruptured in almost exactly the same way in
the two quakes.[3]

The clockwork character of the Parkfield earthquakes convinced some seismologists that fault behavior was a deterministic phenomenon, and that with enough data, the mechanical
"logic" of any seismically active fault could be worked out. The
year after our work at Parkfield, several USGS geophysicists
issued a bold public prediction, with greater than 90 percent
certainty, that a magnitude 6 earthquake would occur there
sometime between 1985 and 1993.[4]

The 1980s gave way to the 1990s. The Berlin Wall was
demolished, and the Cold War ended. The world survived the
Y2K crisis, only to face the horrors of September 11, 2001.
But still no earthquake happened at Parkfield. Finally, in late
September 2004, the expected magnitude 6 event did occur,
eleven years past the end of the prediction window, sixteen
years beyond the twenty-two-year recurrence interval, and

two days shy of the forty-ninth anniversary of James Dean's death.

By this time, a new paradigm in earthquake seismology was emerging, both in the United States and in Japan, which, as a country near the boundaries of four different tectonic plates, had invested massively in earthquake prediction research, to little avail. Predictions had been issued for events that never occurred, while devastating quakes, including one in Kobe, Japan, in 1995 that claimed more than six thousand lives, had not been anticipated. The growing consensus was that earthquakes should be viewed as probabilistic phenomena, inherently unpredictable.

This view gained further traction in the aftermath of the giant M9 subduction earthquakes in Banda Aceh, Indonesia, in 2004, and Tohoku, Japan, in 2011, both of which triggered immense tsunamis. Waveform analysis of these events suggests that even such large megathrust earthquakes do not seem to "know" at their outset whether they will end as modest slip events or develop into M9 monsters.[5] Instead, the eventual size is determined by a cascade of processes along the fault that unfold as the rupture is propagating. Measuring the static properties of fault rocks between earthquakes has little relevance for prediction if the ultimate nature of the quake is dictated mainly by dynamic processes that occur during the event itself.

The 2011 Tohoku earthquake also dispelled notions that certain fault segments always fail in similar-size "characteristic" events. Japanese seismologists had been confident that the segment of the subduction zone that slipped in 2011 would generate earthquakes no larger than magnitude 8, and this underestimate had tragic consequences; tsunami walls and evacuation plans were designed to withstand powerful magnitude 8-to-8.5 earthquakes, but not an M9 event.

For the generation of seismologists who were optimistic about earthquake prediction, accepting the intrinsic capriciousness of earthquakes has not been easy—analogous, perhaps, to spending one's life trying to crack a coded message only to find that the symbols are meaningless. But abandoning simple mechanistic models of fault zones has led to a subtler understanding of their behavior. Today, most geoscientists believe that the best way to protect the public from seismic hazards is through education, effective communication of risk, and advocating for building codes that require earthquake-resistant construction. We can say with absolute certainty that future earthquakes will provide tough tutorials that will further humble and enlighten us.

In the decades since my USGS internship, the narrative around the eruption of Mount Saint Helens has also changed dramatically. The story is no longer one of death and devastation but instead of survival and resilience. Within ten years of the eruption, the barren blast zone had become a natural biological laboratory for the study of ecosystem recovery and succession. The volcanic ash, rich in potassium and phosphorous, replenished soils and provided a rich substrate for new life. At the global scale, the sticky and often explosive magmas generated by water-assisted melting at subduction zones have built the land in a larger and more fundamental sense: Over time, volcanoes like Saint Helens create new continental crust—nonbasaltic, high-silica, un-subductable rock that will remain at Earth's surface over the long term. No other planet has concocted continents.

Similarly, Earth's distinctive habit of subduction—though responsible for both the deadliest earthquakes and most catastrophic eruptions in human history—is essential to Earth's long-term habitability. Crucially, subduction is a mechanism by which the interior and exterior of the planet—mantle, crust, atmosphere, and hydrosphere—remain in communication. Subducting slabs carry ocean water and atmospheric carbon dioxide into the mantle, where these volatiles reside for tens of millions of years before being exhaled again by volcanoes, in an ultraslow-motion planetary-scale respiratory cycle.

Recent studies estimate that Earth's mantle contains between one and three ocean-volumes of water,[6] like an untapped savings account assuring the planet's future well-being. The water in the mantle is not just biding its time there, however. Besides helping to extract continental crust in subduction zones, it has another, subtler role to play in the tectonic system: water reduces the stiffness of mantle rock, allowing the mantle to churn as a flowing solid in that lava lamp–like process called thermal convection. Warm mantle rock expands and rises (and undergoes decompression melting at midocean ridges) then cools, contracts, and sinks at subduction zones. This incessant convective overturning is what powers plate tectonics.

In the absence of water, mantle rocks would probably be too rigid to convect; after 4.5 billion years of gradual cooling, Earth's interior is not warm enough to allow dry mantle rock to flow. A rigid, static mantle means planetary death, with no volcanoes to exhale gases and replenish the atmosphere. This was the fate, billions of years ago, of Mercury, the moon, and Mars. But on Earth, ingeniously, subduction makes the tectonic system self-sustaining, by keeping the mantle hydrated and ceaselessly roiling. It's not just that the Earth *has* lots of water but that its water cycles continuously through all parts of the planet.

To humans, volcanism and seismicity seem merely de-structive, but this perception reflects our short window of observation and the naive conviction that Earth "should" be static. In the logic of Earth, there is no such thing as a "nat-ural disaster." Volcanic eruptions and tectonic upheavals are in fact manifestations of the planet's extraordinary capacity to sustain and renew itself. In other words, to be an Earthling means learning to live with the inevitability of cataclysm.

Volcanology is a far more sophisticated and interdisciplin-ary science than it was when I struggled to set up a reflector in the darkness on the edge of the Long Valley caldera. No-tably, we now have a clearer understanding of the complex effects of volcanism on climate over a range of timescales. We know that the eruption of flood basalts and the associ-ated emission of carbon dioxide led to extreme warming (and several mass extinctions) in the geologic past. In contrast, eruptions of high-silica, subduction-related volcanoes can be explosive enough to send ash and sulfur dioxide particles into the stratosphere, where they act like tiny mirrors that reflect sunlight back to space and cause sudden, short-lived cooling.

This was a surprising lesson of the 1991 eruption of Mount Pinatubo, which for two years temporarily reversed the steady anthropogenic greenhouse-gas warming that was already more than evident to geoscientists. Unfortunately, this observation has been seized upon by people outside the geosciences as a silver bullet solution to our climate woes. Injecting a Pinatubo-equivalent mass of sulfur particles into the stratosphere every year for decades to mask (but not reverse) the effects of greenhouse gases would be ever so much more convenient than having to change our lifestyles in any way. But there are a few hitches. For example, once begun, injections would have to continue to avoid a large spike in warming. Meanwhile, ocean acidification would continue unabated as carbon dioxide continued to climb. In a terrible irony, some solar energy technologies would be made less efficient, undercutting our ability to wean ourselves from fossil fuels. And regional weather patterns—including the Asian monsoon—could be affected in ways that could trigger global food crises and geopolitical instability.

Apart from uncertainties about the wisdom of mimicking large volcanic eruptions, some other fundamental questions in volcanology remain unanswered. In the specific case of Long Valley, we don't actually know why a giant volcano is there in the first place. It's neither above a subduction zone like Mount Saint Helens nor at the site of an obvious "hotspot," or rising mantle plume, like Yellowstone. The most plausible hypothesis is that deep beneath Long Valley, hot mantle rock is bulging upward into a thin, weak spot in the crust and melting as a result of decompression. This "primary" melt—likely basaltic—may then pool underneath the overlying continental crust, and partly melt it, generating high-silica, or rhyolitic, "secondary" melts that are dangerously viscous and explosive.

A more general question is how large bodies of stiff, sticky rhyolitic magma become mobilized in supervolcano eruptions like the one that produced Long Valley's Bishop Tuff or the immensely thick golden tuffs that give Yellowstone its name. In both places, hot springs, earthquakes, and geophysical evidence point to the presence of bodies of semimolten rock in the subsurface. But is an eruption likely at either site? And how much warning would there be before one?

New high-resolution electron microscopes have made it possible to discern growth patterns in tiny crystals in tuffs from past caldera-forming eruptions, and these crystals have some unsettling memories to share.[7] They seem to have resided quietly in hot crystal slurries beneath their respective volcanoes for tens of thousands of years. Then, very abruptly—perhaps in a matter of weeks or months—these crystal mushes were intruded from below by batches of hotter, less-viscous basaltic melt, destabilizing them and triggering gigantic eruptions.

These new insights from tough old tuffs mean that the prologue to a volcanic eruption doesn't necessarily scale with the size of the eruption—that is, there may be no more time to prepare for a supervolcano event like the one that formed Long Valley than there was for the eruption of Mount Saint Helens in 1980, even though the scale and logistics of evacuation would be far more daunting. This is undeniably sobering—but makes our work installing survey stations at Long Valley relevant all these years later. No eruption did happen there in the aftermath of the 1983 earthquake swarm, and the real estate market eventually rebounded. But our first laser measurements of the dimensions of the caldera established a baseline for ground deformation monitoring that continues today. For me personally, those perilous nighttime outings on mountain roads ultimately strengthened my scientific self-reliance. At the same time, working on the rim of a great volcano left me deeply humbled at Earth's formidable powers.

After decades of teaching, I've found that supervolcano

horror stories are compelling enough to capture the attention of even the reluctant students who lurk in the back row of introductory geology classes. They ask how likely another Yellowstone eruption is and whether I lie awake at night worrying about it. I admit that such an event is possible, but not on my list of insomnia-inducing anxieties.

These days, I find contemplating violent geologic change less terrifying than confronting humankind's obstinate *refusal* to change, despite obvious signs of a looming climate crisis. I worry about powerful people with shallow understanding of how Earth works endorsing the idea that we could "engineer" the atmosphere without a torrent of unintended consequences. I worry at the dangerous delusion that nature is simple and mechanistic and is now under our control.

I've now survived enough doomsday predictions, public and personal, to realize that the visions of apocalypse that loom largest in our imaginations aren't usually the ones that come to pass. Instead, it's almost always our own obdurate, self-defeating tendencies that do us in. Meanwhile, Earth makes no distinction between creation and destruction.

DIAMICTITE

By Svalbard standards, it's a fairly nice summer day. The temperature is about 40°F, and based on the ruched look of the surface of the fjord, I estimate the wind to be around twenty knots—what meteorologists call a "fresh breeze." It's raining lightly. We've been walking for almost an hour, and I'm getting warm enough to think it might be worth the trouble of taking off my long underwear. Sweating now means feeling chilled later. One of my companions, a graduate student new to the Arctic and not yet at peace with it, curses the weather with disproportionate profanity. I feel a strange need to come to its defense. When there's no wind, fog can settle in—at least we can see the rocky cliffs that are our destination for the day, and that's something to be grateful for. I suspect that his bad temper is related in part to his insistence on carrying the heavy rifle we have for defense against polar bears—an old wooden bolt-action "thirty-aught-six." We pause for a rest, eat a bit of chocolate, and notice three reindeer not far away, peering at us in a quizzical, nearsighted way that makes us all laugh. I repeat my offer to be the rifle-bearer for a while, and this time my stubborn comrade agrees. Everyone feels better, and we resume our journey.

This is my third summer of fieldwork in Wedel Jarlsberg Land, along the broad Bellsund Fjord on the west coast of

Spitsbergen, the largest island in the Svalbard archipelago. "Svalbard" is an old Norse name that has survived from Viking times, meaning "cold edge." "Spitsbergen," bestowed by sixteenth-century Dutch whalers, is equally apt: "pointed mountains." In three years, I've come to know the moods of this elemental landscape and accept the austere hospitality it offers. When the weather is tolerable, there is no time to waste; the grant supporting my PhD research at the University of Wisconsin is soon to expire. This is my last opportunity to gather observations and samples for my dissertation, whose main product will be a detailed geologic map covering more than one hundred square miles of rugged topography. At the urging of my well-meaning advisor, I have already written a master's thesis based on my first season of observations—a fallback plan in case I don't finish the doctorate. He's skeptical, I suspect, because I've recently married a fellow graduate student; I've overheard faculty members grumble that women students are likely to drop out once they've found husbands. In reality, I am already having misgivings about the marriage, and I welcome all the challenges of Arctic fieldwork, which keep me from brooding too much on my personal life.

Creating a geologic map in such terrain is intellectually and physically demanding. It requires stamina, mental toughness, and the capacity to think in not just three but four dimensions, including time. By the middle 1980s, however, the field of geology was changing, and a new bedrock map, whatever its heroic backstory, was not itself enough to merit publication in any of the leading peer-reviewed professional journals. To get my work in print—essential if I want to continue in academe—I will need to develop a coherent account of the tectonic evolution of the region. This year, I've also been charged with mentoring two master's students in their own projects. The Arctic summer is short, and whining about the weather isn't helpful. I've learned that it's important in the first days of a field season to establish a sense of esprit de corps, a shared habit of cheerful stoicism that will help buoy everyone up under difficult conditions. The conditions today are not difficult, and I'm a bit worried about how my companions will cope when the weather really does get bad. To get our jobs done here, we will need to be receptive to what the rocks have to say, and that can't happen if one has an antagonistic relationship with the landscape. Deliberately changing the subject, I point out the names of the topographic features in our study area.

We're camped in a mossy valley called Chamberlindalen, in honor of the polymathic geologist T. C. Chamberlin, who found evidence for multiple episodes of glaciation in the Great Lakes region, articulated important principles of groundwater hydrology, and even developed an early theory of solar system formation. Chamberlin was president of the University of Wisconsin from 1887 to 1892, which makes the valley seem even more like home. Not far away is Kapp, or Cape, Lyell, named for the eminent Sir Charles Lyell, whose 1830 treatise *Principles of Geology* was the foundational work for the new science of Earth. So we're in good geologic company here.

The entire west side of Chamberlindalen is a massive wall of a single, unusual, sedimentary rock type: diamictite, formally known here as the Kapp Lyell sequence. Like conglomerate, sandstone, or shale, diamictite is a "clastic" sedimentary rock—i.e., one made of physical particles as opposed to "chemical" sedimentary precipitated from water, like limestone or rock salt. Most clastic sedimentary rocks have a fairly uniform grain size that reflects the velocity of the water or wind that deposited them: a conglomerate made of pebbles or cobbles points to a fast-flowing current, while a fine, clayey shale indicates quiet water conditions on a lake bed or seafloor far from shore. But diamictite (literally "mixed rock") is an exception—a strange combination of large stones embedded in a fine-grained matrix.

We can imagine at least two processes that could leave such a deposit. A viscous mud flow could entrain cobbles and boulders and end up as a mélange of clay and stones, but mud flows would be rare in the rock record. For one thing, they tend to be localized in extent, and also, since they occur on hillsides, they are likely to be eroded away—not buried, preserved, and hardened into bedrock. A second, more plausible scenario is suggested by the icy landscape around us in Chamberlindalen: these diamictites might be ancient glacial deposits. On our way up this valley, we crossed several old moraines—jumbles of boulders, cobbles, pebbles, and fine-grained "rock flour" left by the glacier that filled Chamberlindalen at the height of the last ice age. If that mixed material, which geologists call "till," were one day covered by other sediments and consolidated into rock, it would form diamictite. Indeed, many diamictites are "tillites"— "petrified" till.

The stones in the Kapp Lyell diamictites range in size from pea-size pebbles to boulders bigger than watermelons. Both comparisons feel like anachronisms, however, since these rocks, at about 650 million years old, predate the earliest land plants of any kind by 200 million years and flowering plants by 500 million years. These clasts also represent an array of rock types: most are carbonate rocks, primarily dolostone, but there are also chunks of quartzite and some igneous rock types. The varied sizes and compositions of the embedded stones are consistent with their having been picked up indiscriminately by a glacier as it scoured an ancient land surface.

But the interpretation of these rocks as an ancient till deposited directly by glacial ice doesn't quite fit because the fine-grained matrix of these diamictites is very distinctly *layered*—clearly not the work of an advancing sheet of ice, which, like a bulldozer piling dirt in front of its blade, would leave behind an unstratified heap of sediment. The layers in the Kapp Lyell diamictite are thin and uniform, about a half inch thick, and come in couplets: layers of very fine sand, alternating with thin films of darker clay. These look very much like "varves"—a distinctive type of laminated sediment that accumulates in high-latitude lakes or fjords. The term comes from a Swedish word meaning "turn" or "cycle"; "solstice" in Swedish is *solvarv*—the turning of the sun. A single varve, like early and late wood of a tree ring, is the record of one year: a coarser summer layer represents the input of flowing streams, and a thinner winter layer reflects the slow rain of fine suspended particles when ice covers the surface of the water. The uniformity of varves points to quiet conditions at the bottom of a deep body of water.

So how could large rocks become interbedded with such finely laminated sediment? There is only one, very specific, explanation: icebergs with cobbles and boulders frozen into them calved off their parent glacier, floated out into an otherwise placid waterbody such as a glacial lake or fjord, and then broke up, dropping their stony freight—*plunk, plunk*—to the lake bed or seafloor, like smugglers jettisoning illicit goods. Looking more closely, we note an asymmetric pattern: The stones disrupt the bedding on one side, but the layers drape over them on their other side. This is clear evidence that they were in fact "dropstones" carried out into the water by floating ice. It's also an indicator of the original direction in this rock sequence, which has been tilted and folded by tectonic forces.

It's not too difficult to imagine the process that formed these rocks, since out in the fjord behind us there is a modern flotilla of icebergs recently detached from one of the glaciers that empties into this branch of Bellsund. Those icebergs no doubt carry chunks of rock that will plummet to the muddy floor of the fjord, forming a new generation of diamictite. We laugh at the thought that some of those chunks may even be pieces of the ancient diamictite, reenacting their birth 650 million years later. The rocks of Chamberlindalen predate plants and animals, Pangaea, and the Appalachian-Caledonian mountain belt, but as we look out over the ice-strewn fjord, it feels as if we're witnessing their formation in real time.

Without consciously thinking about it, we are employing two fundamental conceptual practices linked with the geologists whose names have been assigned to this landscape. The first is the principle of uniformitarianism, often summed up in the phrase "the present is the key to the past"—i.e., the idea that in interpreting records of Earth's past we should look at processes happening in the world today. It's just a happy coincidence that in this case we don't need to cast our eyes farther than the adjacent inlet for inspiration. Uniformitarianism was the leitmotif of Charles Lyell's three-volume monograph, the one big idea he strove to inculcate in his readers. He did it so well that for more than a century, uniformitarianism was a rigid kind of orthodoxy that made it taboo to invoke catastrophic events that had not been experienced in human history—e.g., cataclysmic floods, asteroid impacts, or super-volcano eruptions—as explanations for geologic features.

The second cognitive practice is one of T. C. Chamberlin's most enduring contributions to geology: the "Method of Multiple Working Hypotheses." In an 1897 paper[1], he exhorted geologists to adopt the habit of holding in the mind more than one scenario for how rocks, waterbodies, or landforms could have come into being and, without playing favorites, identifying tests for each interpretation until one is shown to be the most likely. Chamberlin's ten-page exposition of this simple idea is a bit overwrought, but there is no doubt that the method of multiple working hypotheses is good practice—not only for interpreting rocks but also in criminal investigations, medical diagnoses, and truth-seeking in any context.

At the time we were mapping the Kapp Lyell diamictites in the mid-1980s, similar rocks—including some true tillites—of about the same age were known from mainland Norway and Scotland, but since these were geographically clustered in the North Atlantic region, there was no Lyellian reason to postulate that they represented anything more than a moderate ice age in latest Precambrian time. In the next two decades, glacial diamictites from this same period began to be found around the world, including in places that had been near the equator 650 million years ago—evidence of an extreme, non-Lyellian ancient ice age, of much greater duration and severity than the much more recent Pleistocene. This "Snowball Earth" interval and its effects on ocean chemistry are now thought to have created conditions that fueled the explosive emergence of animal life in the Cambrian.

I would return to Chamberlindalen twenty years later to understand what the Kapp Lyell diamictites could reveal about that inflection point in Earth's history. But as a graduate student, I was single-mindedly focused on what they would tell me about the tectonic construction of the northern Caledonides. To me, the dropstone clasts were most useful for quantifying the nature of the deformation, or "strain," that the rocks had experienced as the mountains formed.

Here was the logic: Although no individual dropstone would have been a perfect sphere in its original state—some might have been flattish, others ovoid—on average, they would have been approximately spherical. Also, they would not have had any sort of original preferred orientation because they were deposited by falling to the muddy seafloor, not by a directional current. Stones deposited by moving water tend to be tilted in the upstream direction and stacked on each other like tiles on a roof, but stones falling through standing water from deteriorating icebergs would land in the mud however they may. Thus, any statistical deviation from sphericity or tendency for alignment would reflect later deformation of the sedimentary rocks during mountain building. So that day in Chamberlindalen, as most days that summer, I was seeking out places with good three-dimensional exposures of the diamictites where I could measure the dimensions and orientations of hundreds of dropstones.

Sometimes, with my ruler in hand, I would realize how ridiculous this task would appear to someone who might happen by—but, of course, no one else was out there to happen by. There was a rookery of little auks at the top of the cliffs whose guano made Chamberlindalen so green and mossy, and it always sounded like those birds were having a raucous party. I found their jovial company comforting and didn't blame them for laughing.

My field, structural geology, is concerned with documenting the geometry of deformed rocks as a first step toward understanding the processes that occur during mountain building. We can't get inside mountains like the Himalaya that are still forming, but we can walk around in the interiors of deeply eroded mountain belts and then attempt to reconstruct the sequence of events that must have happened as they rose. Measuring pebbles and cobbles was a tidy, satisfying data-generating exercise that helped us feel like we were making progress toward fathoming the large-scale architecture of the mountains, a daunting goal both conceptually and practically. This was especially difficult in the Kapp Lyell diamictites, which are thousands of feet thick but lack distinctive "marker beds" that can be readily traced to map folds and faults.

Svalbard was a far more isolated place in the 1980s than it is today. We had no means of communication with anyone outside our camp other than the two-way radio we used for a nightly check-in with the Norwegian Polar Institute, which provided our logistical support. We needed to measure out a span of antenna wire to a precise integer multiple of the wavelength corresponding to the radio frequency the Institute used—or be cut off entirely from the world. If someone in our group had been seriously hurt and needed help at a time other than the designated radio hour, there was an emergency frequency that was usually monitored by someone in the mining settlement of Longyearbyen two fjords away, but using that would have required reconfiguring the antenna under stressful circumstances. And if we did manage to convey the message that we needed help, it would still have taken ten hours for a boat and more than an hour even for a helicopter to reach us (at great expense). In modern life, layers of technological infrastructure buffer us from the physical underpinnings of reality; in our primitive, isolated camp we felt the raw physicality of everything, including the dimensions of invisible radio waves.

Our daily existence was complicated in the way camping always is; without plumbing, electricity, or furniture, tasks like preparing food, washing clothes, and keeping oneself reasonably clean are just more elaborate. Our work was physically demanding; each day, we'd hike many miles, climb thousands of feet, and return to camp with backpacks full of rock samples. Yet at the same time, our life was simple and focused—almost monastic. Our single task was to translate and transcribe the story of the mountains.

Although the sun would never set (but seldom *shone*) in our eight-week field season, we maintained a consistent routine to structure our time. Leaving a cozy sleeping bag in the morning was probably the hardest task of the entire day. Being the first to get up, light the tiny Primus stove, fetch water for coffee, and start a batch of oatmeal was an act of moral fortitude. In such a remote place, a field party is a tiny society in which all citizens must share civic responsibilities. Sometimes the hardest work is psychological and not merely practical—maintaining good humor and bonhomie takes effort when people are cold, tired, and homesick.

For my mental equipoise, especially during days when lashing rain trapped us in our tents, it was essential to know I could escape into a good, immersive book. Choosing the right volumes to bring for a two-month field season was as important as packing enough toothpaste. The first time I went to Svalbard, I thought that I'd want to read light, frothy things that were the antithesis of being in a wilderness outpost and brought a selection of P. G. Wodehouse titles and Lord Peter Wimsey mysteries. But I found it hard to care about Bertie Wooster's ridiculous predicaments or the architectural details of English manor houses while lying in a sleeping bag as gale winds pressed tent walls against my face. In subsequent years, I opted for novels by Thomas Hardy, George Eliot, Halldór Laxness, Louise Erdrich, and Annie Proulx in which nature is a force field that shapes the actions of human characters. My escapism required some connection to our reality.

That summer in Chamberlindalen, I tended to be the first in our group of four to emerge from my cocoon, get breakfast started, and lay out supplies for lunch. A daily aggravation was that each night, one particular member of the group would take the only knife that was suitable for bread-cutting to bed with him so that if a polar bear attacked his tent while he was sleeping, he would be able to slash his way out. I trusted that any bear coming for a visit would first run into the trip wires we had set up around our campsite, which would detonate a small explosive device that would wake us all up. The system was well tested because it was not unusual for someone to forget about the wire and set off the *snublebluss* (Norwegian for "stumble flare") after emerging groggily from their tent at night to pee. Retrieving the bread knife was a ritual that marked the official beginning of the day.

We could drink water from a nearby glacial river without boiling it, but not before letting the fine gray "rock flour" settle out of it for at least a day. We washed dishes downstream

from camp by scrubbing them with sand and gravel. Because
I tended to care more about clean cookware than any of my
male compatriots—who were happy to make an evening stew
in a pot with remnants of the morning's porridge—the sedimentary dish-scouring generally fell to me. Warming water
for washing dishes seemed profligate; we had three five-gallon
metal jerry cans of kerosene for the season, and everyone was
concerned about running out.

My fastidiousness and frugality had a price, however. A
week or so after returning from my first summer in Svalbard,
my fingers became painfully swollen, red, and blistered. Fearing that I had a serious skin infection, my doctor, a young
University of Wisconsin medical school resident, prescribed
antibiotics. When, after several days, these had no effect, he
asked a senior colleague to look at my inflamed hands, and
the older doctor immediately recognized my condition as an
affliction that he hadn't seen in decades. With some incredu-
lity he asked, "How on Earth did you get chilblains—in the
summertime?" Daily dish washing in glacial meltwater had
damaged the capillaries in my fingers, causing blood to leak
into the surrounding tissue. It took several months before
my traumatized hands looked normal again. In subsequent
Arctic field seasons, I washed dishes decadently, in warm
water.

Our food was ordered through a Norwegian supplier of
provisions for ships and came in bear-proof plastic boxes that
also served as table and chairs in the cook tent. The boxes
were designated by letter, with *A* boxes containing staples like
coffee, tea, oatmeal, sugar, salt, flatbread, rice, and dry milk;
B's canned fruits and vegetables; *C*'s tinned meats and fish
preserved in many forms; and *D*'s luxury high-calorie items

like butter, honey, preserves, biscuits, baking chocolate, and caramel-colored *gjetost*, Norwegian goat cheese.

Our bread came from the lone bakery on Svalbard, in the town of Longyearbyen, in massive four-foot-long "expedition loaves" that became progressively moldier as the season progressed. We would carve off the green rind until eventually all that remained edible would be a narrow cylinder from the interior of a loaf. To mask the taste and quell our constant hunger, we would butter not only both sides but also the edges of the bread slices and cover them with jam or, when that ran out, plain sugar. The cook tent was warm but small and cramped, and the floor was always muddy—so dropping a piece of well-buttered bread was catastrophic. It seemed that any indoor space soon became grimy while the tundra always felt fresh. On the rare days when the weather was nice enough, we preferred to sit outside, using sun-bleached whale vertebrae from the beach as chairs.

If a Norwegian Polar Institute helicopter had other business in our part of the archipelago, we might be lucky enough to get a few fresh loaves, and maybe even some carrots, cabbage, or potatoes. The crew would climb out of their aircraft like alien visitors from a more advanced civilization—with their freshly laundered jumpsuits, perfectly coiffed hair, news of world events, and piteous glances at our little tent camp. Suddenly, we would realize how grubby our clothes were and how unkempt we must look. Often, just before leaving, the pilot would say "Oh, I almost forgot, there is some mail," knowing full well that letters from home were what we craved even more than bread without mold. Even today, the low, pulsing sound of a helicopter overhead always stirs a subliminal feeling of anticipation in me.

The year we were in Chamberlindalen, we had a surprising number of guests materialize from the sea: Japanese geologists working on the far side of Bellsund who arrived by rubber Zodiac; a charming French couple sailing their yacht—with a cat and baby grand piano—around the world; a Russian fishing vessel whose captain sent a small boat out to bring us on board so he could tell us what a great man Ronald Reagan was. These encounters made it seem that Svalbard was the crossroads of the world. We also had nonhuman callers: a pod of beluga whales who stopped by several times, snorting their salutations, and swimming close enough to shore for us to look one another in the eye.

In my second season in Svalbard, when we were a camp of just three, we had visitors arrive on foot. Our nearest neighbors that summer were forty miles away, at a Polish base station on Hornsund, a major fjord to the south, separated from our study area by two deeply crevassed glaciers and a series of formidable peaks. One evening, not long after we had returned from a long day of fieldwork, we were astonished to see four distant figures coming from the south. We ran like school-age children out to meet them. They turned out to be Poles who spoke very little English, but with whom we were able to communicate through a combination of gestures and drawings as well as my passable German and minimal Russian.

After much hilarious misunderstanding, we learned that they were not scientists but plumbers and electricians who had earned a working vacation at the Polish station. We joked that with us they were truly on vacation, since we had neither plumbing nor electricity. When we expressed amazement at the journey they had made (without commenting on the

tragic state of their footwear or their rudimentary backpacks),
they simply said they wanted to see other parts of Svalbard
(and escape their Communist Party minders) before returning
to Warsaw. My companions and I cooked the best dinner we
possibly could with our ships rations, and despite having no
common language, everyone talked late into the sunlit night.
Finally, exhausted from the sheer exertion of so much unaccustomed social interaction, we all collapsed into our sleeping
bags.

The next morning the Poles needed to make an early start
to get back to the station by the promised time. They show-
ered us with candies and preserves whose labels were small
windows into daily life within the Soviet bloc. We pressed
them to take coffee, sausages, wool socks, and a spare ice axe,
and they set off on their unimaginable trek back over the
mountains and glaciers. The visit felt like the most fundamentally human encounter I'd had in my life—a reenactment
of age-old stories of travelers seeking shelter, a reawakening
of ancient rituals of hospitality encoded deep in our genes.

After the emotional intensity of these visits, it was always
something of a relief to return to our single-minded routine of
fieldwork, steadily filling in the blank map with information.
There was comfort in the evening rituals of oiling our boots
and hand-copying our field notes in case the originals were
lost in a stream or crevasse. Each year, I tried to keep a diary
but found it impossible—the experiences were too intense to
look at directly; writing about them as they happened was like
gazing straight at the sun. But behind all the flat observations
in my orange field notebooks on rock types, orientations of
faults, and dimensions of diamictite clasts are vivid unwritten
chronicles of those days.

The experience of being fully immersed in, and dwarfed by, the object of one's study leads to a shift in consciousness—akin, perhaps, to what writer and philosopher Iris Murdoch called "unselfing," in which the "fat relentless ego" is finally subdued.[2] To Murdoch, unselfing was essential to becoming a moral being. If so, places like Svalbard are academies for moral training; the stern logic of the landscape necessarily redirects one's thoughts outward. About three weeks into each field season, my dreams would change from being set in civilization—often involving anxious scenarios like having forgotten to study for an exam—to taking place on the tundra, where I was a participant, but not the protagonist. The austerity of the Arctic environment, where there is no place for euphemism or artifice, made a permanent impression on my psyche. I developed a deep sense of myself as an Earthling, a hatchling, a foundling. A small creature born to a vast, old, rocky enigmatic planet.

Every day in Svalbard, the land revealed more of its creative experiments with limited materials. Ice on the surfaces of glacial tarns—high mountain lakes—broke into long, tinkling crystalline rods like prisms of an elaborate chandelier. The flat strandplain was covered with remarkably regular circles of gravel-size rock about six feet in diameter, creating giant honeycombs the size of soccer fields. Exactly how such "patterned ground" forms is still not completely known, but it seems to be related to freeze-thaw cycles that churn the soil above permafrost zones. Gently sloping hillsides were lumpy with green tuffets arranged in clusters that resembled half-scale Hobbit villages. These were charming but made for difficult walking. We became experts at identifying different tuffet subvarieties, able to distinguish, at a glance, simple

moss-covered boulders from grass-topped mounds that had
elevated themselves above boot-sucking, marshy ground.

The palette of the tundra is subtle, almost grayscale. One
autumn evening after my first field season in Svalbard, I invited friends over to share photographs I had taken. I bor-
rowed a slide projector from the geology department, tacked
a bedsheet on the wall of my apartment, made some popcorn,
and we settled in for a show. After a few minutes, someone
asked why I had used black-and-white film—but all my photos were taken with glorious Kodachrome. It simply takes
time for the eye to become attuned to the understated hues
of the Arctic landscape. And the colors change quietly over
the course of the short summer, from muted versions of jade
and periwinkle to dusky ochre and dun.

The few land animals native to Svalbard—polar bears,
reindeer, fox, and various species of birds—have all adopted
the colors of ice and rock. Arctic ptarmigans, in particular,
are so well camouflaged that we would nearly step on fledg-
lings following in a line behind their mothers, who looked
like waddling stones. In contrast, if we got anywhere near the
broods of Arctic terns, whose nests blended invisibly into the
rocky ground, we were in for an attack reminiscent of scenes
from Alfred Hitchcock's horror classic *The Birds*. I simultaneously detest and respect terns; every year, they make the
extraordinary trip from the Arctic to the Antarctic and back,
in pursuit of endless polar summers.

Late in the Svalbard summer, the skies become clearer but
the sun circles perceptibly lower and lower each day. By the
end of August in Chamberlindalen, the surrounding peaks
cast long shadows into the valley at all hours, and we began
to feel like light-starved plants. One cloudless day, the radiant,

The diamictite outcrop that revealed two distinct episodes of deformation.
Black limestone clasts were first flattened, then folded.

sunlit slopes high on the mountainside beckoned to us like an
Arctic Shangri-La. We decided to climb up to the illuminated
cliffs—and found diamictite outcrops that revealed the tectonic story of the area.

At this site, many of the dropstones were a distinctive
black, organic-rich limestone. The dark color of such limestones usually comes from the presence of carbon in the form
of graphite, which makes the rock especially weak and able to
flow in the solid state. Unlike the other clasts that I had been
measuring, which were ellipsoids ranging from the shape of
rugby balls to curling stones, these limestone clasts were crin-
kled trains of M's or W's. Those letters were the cryptographic
key to decoding the history of the mountains.

Here is the message they conveyed. The zigzag limestone
clasts must first have been flattened into pancakes during an

early stage of mountain building, probably from the weight of rocks stacked up on each other by faulting. Then, in a second episode of deformation—perhaps during the final collision among continents that formed Pangaea—they were accordioned into tight folds. The other, less-sensitive types of clasts—the rugby balls and curling stones—had simply recorded the cumulative effect of both stages. It was a Rosetta Stone moment. We now realized that the complex large-scale structures we had been struggling to interpret had been produced by two distinct stages of mountain building. We had been trying to read the rocks as if they told a story with one plotline when in fact there were two entangled narratives. It was hard not to feel, on those spotlighted slopes, that the mountains had waited until we were ready to receive this new information and had chosen that moment to unveil it.

September arrived, the tundra donned its fall colors, and the field season came to an end. Transitioning from a fully functioning camp to one ready to be picked up was logistically complex and psychologically taxing. We were in radio contact with the Norwegian Polar Institute ship due to retrieve us, but there were several other parties to pick up first, and the ship's arrival could not be predicted with confidence. It could happen that afternoon, or sometime in the middle of the night, or possibly the next morning or later. Depending on how choppy the waters of the fjord were, we would be brought to the ship either by rubber boat or helicopter.

Precious rock samples had to be carefully wrapped to protect them from rough baggage handlers and redundantly labeled in case they were opened by suspicious customs inspectors. Tents had to be taken down, even if it was raining. Food and cooking supplies had to be packed away—but also accessible in case we were waiting for hours. Rifles had to be at hand until the very last minute, when they were to be unloaded. And our minds became divided between our surroundings and the prospect that we would soon be on the way home. Our agenda was being set by humans rather than nature again. We began to feel agitated, impatient.

Once we were aboard the ship, and the mountains that we knew so intimately slipped from view, the process of re-selfing inevitably began. Taking a shower for the first time in two months is an extraordinary experience. One's own naked body is unfamiliar, and standing under a stream of hot water, scrubbing off grime and dead skin is something close to being born again. In our gleaming state, donning clothes not prescribed by the weather outside, we eased back into human society. The voyage back to mainland Norway was festive; all the scientific field parties were full of stories of the summer's adventures—close calls on glaciers, mishaps in rubber boats, encounters with polar bears.

After the ship landed in the gritty Arctic port of Bodø, we spent a day as dockworkers helping unload the hold and stowing equipment for the next year. Then, staggering under the weight of all our gear and rock specimens, we boarded a train for the eighteen-hour journey to Oslo. Summer in Norway's capital is always sublime, but after our ascetic existence on the tundra, sensory experiences of urban life were almost transcendental. Simply walking down smooth sidewalks felt luxurious after a summer of stumbling over rocky ground. Oslo is not exactly tropical, but to our eyes it seemed riotously green; we marveled at how plants could grow so extravagantly. We almost wept at the joy of eating fresh produce and drinking beer.

We visited the offices of the Polar Institute to share our findings with some of the senior geologists, who included some heroes of the Norwegian resistance during World War II, then well into their sixties but still spry and tough. Our primary contact, however, was Yoshihide Ohta, a Japanese geologist who had come to Norway for a short-term position twenty years earlier but ended up emigrating from Japan with his family. Ohta's knowledge of Svalbard's complex "basement" rocks—the oldest and most deformed ones—was encyclopedic. In addition to being a brilliant geologist, he was an artist, epicure, and philosopher. He would make astonishingly accurate sketches of rugged mountainsides in a few minutes, then annotate them with geological observations. He had a round belly and wry sense of humor and didn't mind when people called him Buddha or Yoda. Ohta-san would become an important mentor to me in future years, teaching me to broaden the scope of my thinking to regional tectonic scales—and how to make a proper pot of rice on a tiny camp stove. But now, after a few glorious late-summer days in Oslo, it was time to fly back to the United States and face the new academic year, which had already begun.

The psychological stresses of graduate school, which had receded while we were in Svalbard, closed back in quickly, intractable as polar bears. There were plenty of relentless egos prowling the academic halls. In the absence of a formal system for reporting sexual harassment, our small cohort of women graduate students kept one another informed about which faculty members to avoid when one was working in the labs at night or on weekends. Only a few of them were predatory, but we suspected that even the more supportive professors would not necessarily take actions against their colleagues if we filed a complaint. So we were on our own.

I often thought longingly about life on the tundra, which seemed simple and cozy by comparison. The rest of our department was not much interested in the enormity of the experiences we'd just had. At that time, geology was redefining itself as a more rigorous, quantitative science, and there was a generational battle unfolding between an old guard whose work was based almost entirely on field observations and a brash new breed of geologists who were primarily lab-centered instrumentalists and numerical modelers. Merely surviving for two months on the tundra wasn't cutting-edge science.

I had a dawning awareness that my advisor, a kind, unpretentious man who was a pioneer in the geology of Antarctica, was on the "wrong" side of this generational divide and that by association, so was I. If I wanted to have a viable academic career, I would have to claim a place among the rising new cohort of geoscientists by teaching myself new lab-based analytical methods. I would need not only to distance myself from old-school field geology but also, I soon realized, to avoid speaking about field experiences as transcendent spiritual epiphanies. It would be dangerous to talk about adventures in tuffet villages or eye contact with beluga whales, insights you've had about the human soul, or the conviction that mountains were speaking to you.

I was well aware of the further challenges I would face in applying for academic jobs. Small, female, and implausibly young (finishing my PhD at age twenty-four), I just didn't fit people's preconceptions of what a geologist looked like.

Moreover, my dissertation focused on a place beyond the edges of many world maps, far from the trendy centers of structural geology at the time: the Appalachians, Rockies, and Basin and Range region of the United States. So in that last year of grad school, I forged a no-nonsense regimen to bootstrap myself into a competitive position. Although I had already completed all coursework required for the doctorate, I sat in on geochemistry courses that my advisor hadn't thought were important. I sought mentors at other universities and worked my way through an ambitious reading list of the latest papers in structural geology. I found ways I could apply my experience in geophysics and graduate coursework in rock mechanics to make my work more quantitative. In short, I became insufferably single-minded.

And then there was my marriage, which had been volatile from the start—an unstable compound of opposing personalities—and was now heading toward rupture. When we met as first-year graduate students, my husband's "don't worry, be happy" worldview had been an irresistible counterpoint to my default setting of underlying anxiety. I also loved the story of how his immigrant parents, who had grown up in different hemispheres, met in North Dakota, of all places, and how tamales and pierogies sat side by side on the table at their family gatherings. In the months before we got married, doubts kept bubbling up in my mind, but the juggernaut of wedding planning kept propelling us forward.

The two of us shared an appetite for outdoor adventure, though it vexed him that I was faster on cross-country skis. He benefited from my organizational skills and frugal housekeeping, and I learned from him to enjoy good beer and the occasional joint. But as the realities of adulthood loomed larger, our differences became flashpoints for conflict. We had incompatible views about money, time management, and what we would do after graduate school. My obsessive work habits steadily undermined our relationship. He bought a fancy new bicycle without talking with me first, draining our meager savings account. In the four years it took me to finish a master's and PhD, he had not yet completed his master's thesis, but he'd certainly had a lot more fun.

A year after the summer in Chamberlindalen, I had my doctorate in hand and was ready to look for academic jobs. My husband, however, expected not only that I would wait for him to finish his PhD at a university where there were no prospects for me but also that we would then try to find positions in the same place. That could take five or six years, by which time I would no longer be a viable candidate for a faculty position. I felt trapped, and he felt wronged. We split, not so much with bitterness as with embarrassment at our mutual misjudgment, at not knowing ourselves well enough.

In later summers, as a young faculty member, I would work in other parts of Svalbard. In 1989, Ohta invited me to be part of a Norwegian Polar Institute project with Polish geologists at the station in Hornsund (whence our visiting plumbers had come). That was the year when the trade union Solidarity successfully pressured the Communist government into negotiations, an important step toward the eventual end of Soviet rule. Through the eyes of my Polish collaborators, I felt that I vicariously experienced those momentous political changes, the once unimaginable end of the Cold War.

At the same time the Cold War thawed, geologists were discovering evidence for an extreme deep freeze, the most

extreme climate excursion in the planet's history: Snowball
Earth. The Kapp Lyell diamictites, which had once seemed
an oddity in a remote corner of the world, were now at the
center of a global story.

Geologists concur that the Snowball Earth interval—now
formally named the Cryogenian ("cold-born") Period of the
Neoproterozoic Era—was a deep ice age. It lasted more than
80 million years, from 720 to 635 million years ago—more
than the time that separates us from the dinosaurs, and about
thirty times longer than the duration of the recent Pleistocene
ice age. It was also far more severe. Diamictite deposits of
Cryogenian age have been found all around the world, and
many are varved marine deposits with dropstones, like the
Kapp Lyell sequence. Reconstructions of tectonic plate con-
figurations for this time in Earth's history show that some
of these "glaciomarine" diamictites formed near the equator.
This means that during the Cryogenian, ice covered not only
the poles and midlatitudes but also reached sea level in the
tropics.[3]

Still, some fundamental questions remain about the Snowball Earth interval. Was the world truly a "snowball," with
oceans entirely frozen, or more of a "slushball," with areas of
open seas, like the perennially ice-free regions called polynyas
in the modern Arctic? Also, how did the Cryogenian deep
freeze end: gradually, as a result of steady accumulation of
volcanic carbon dioxide, or more abruptly, perhaps through
the explosive release of biogenic methane from seafloor deposits called gas hydrates? These questions had not yet been
formulated when I was working on my PhD and focusing
on what the Kapp Lyell rocks could tell me about their adult

tectonic experiences. I yearned to go back and ask them about their childhood.

Twenty-one years after my summer of measuring drop-stones, I was back in Chamberlindalen at the invitation of Polish colleagues—now ardent free-market capitalists—to help develop a unified understanding of the geologic evolution of the area from Hornsund to Bellsund.

The generational battles in geology that had loomed large when I was in graduate school were long over; the discipline had emerged as a more sophisticated science that embraced a wider range of investigative tools, including field-based study.

Two decades of life had changed me as well. I was well established in my career, had remarried, become a mother. I had learned to slow down, set work aside, savor the moment. Motherhood even made me more forgiving of Arctic terns: I now understood their ferocious defense of their nests.

I didn't expect Chamberlindalen to be much altered, how-ever, and was stunned to see how much ice had disappeared since we had last been there. Rocks that had been blanketed by glaciers in 1986 were now laid bare; the disappearance of modern ice had, ironically, provided access to a more complete account of the ancient Cryogenian ice age. My eagerness to study these newly available records was mixed with shameful grief at how radically the landscape had been altered in such a short time by human hubris. Even the little auks' clifftop parties seemed more subdued.

The new exposures made it possible to make a bet-ter estimate of the immense thickness of the Kapp Lyell diamictites—on the order of ten thousand feet. In revisiting these rocks, I also saw a pattern I hadn't fully grasped before.

The stratigraphically lowest—oldest—part of the Kapp Lyell diamictites has very few dropstones and, in fact, seems to be continuous with an underlying black shale, or mudstone. The appearance of the first dropstones coincides with the first varve-like layers. Continuing farther up in the sequence and forward in time, dropstones become more and more abundant. Then, quite abruptly, the varves disappear, and for an interval of several hundreds of feet, the rocks are a disorganized blend of fine and coarse material. These, in turn, give way to more orderly layers sorted by size: cobbles and boulders at the base, grading upward into pebbles, grit, and progressively finer sand.

It took me some time to understand the significance of this sequence—from mudstone, to varved layers with increasing numbers of dropstones, to chaotic jumble, to size-sorted beds. Curled in my sleeping bag one night, I had a half-conscious dream about what the rocks were saying. I realized that the rocks recorded a great acceleration, like a musical piece that starts very slowly and grows continuously faster.

The mudstones below the diamictites, themselves thousands of feet thick, probably represented the depths of the Cryogenian ice age—a silent frozen world where almost no sediment was coming into the water and deposition occurred at a larghissimo pace. The varved layers represent an awakening, evidence that there were seasons again, and enough thawing for sand-size sediment to be delivered in yearly allotments to the seafloor, but the tempo is still adagio. Dropstones mean that icebergs are breaking off into open water, and they begin to accumulate in greater numbers—andante, allegro, vivace. The jumbled, disorganized interval that comes next suggests the launching of whole armadas of icebergs, perhaps the collapse of an entire ice shelf, happening so quickly that no varves accumulate; the pace is now presto. Finally, the size-sorted beds at the top of the sequence record the reworking of these unstable heaps of ice-rafted material by turbulent submarine avalanches.[4]

In other words, the Kapp Lyell diamictites record not the peak of Snowball Earth but instead its dramatic end—the collapse of a great ice sheet, eerily similar to scenarios glaciologists fear could occur in West Antarctica today. It is almost too perfect an irony that rocks named for Charles Lyell, the patron saint of uniformitarianism, in fact represent a genuinely catastrophic event, a singular moment when the world changed.

Standing at an outcrop where an earlier version of myself had measured the dimensions of dropstones two decades earlier, I had a vertiginous feeling of being unmoored in time, revisiting not only my own past but also seeing in the rocks a nightmarish glimpse of the future. And I realized that at any given time, rocks reveal to us only what we are ready to understand. Their meanings are always multiple, and our hypotheses usually naive.

TURBIDITE

High above a glacial valley on a cold, sunny August day on northern Ellesmere Island in Arctic Canada, I'm trying to get oriented in space and time. Two years have passed since the summer in Chamberlindalen, and it's glorious to be back in the far north. My field assistant, an undergraduate geology student from Toronto, is standing stunned at the view out over the Ward Hunt ice shelf and, beyond that, the endless frozen sea. I know where we are on the map: latitude 82°00' N, longitude 83°06' W. But this far north, compass directions and time zones become fluid. We are in fact north of the north magnetic pole, which at this time lies three hundred miles to the southwest, on Ellef Ringnes Island, named for a Norwegian beer brewer who sponsored an 1899 expedition to chart then-unknown parts of the high Arctic. The magnetic pole has been staggering drunkenly around this part of the Canadian Arctic Archipelago for a few hundred years, and using a navigational compass here is a tricky business.

Earth's magnetic field, generated by the roiling motions of molten iron in the outer core, is a scintillating, shape-shifting marvel, an essential planetary amenity that most people rarely even think about. The magnetic field shields the Earth not only from the rapacious solar wind, which could strip away the atmosphere over time, but also from cosmic rays that dart in from deep space with enough energy to damage living cells. At low latitudes, the field seems relatively stable, but closer to the poles, its dynamic nature is obvious.

Here in the vicinity of the magnetic pole, the local north direction can vary over timescales of days. Also, the lines of magnetic force are nearly vertical, so we have to counterweight our compass needles with copper wire to prevent them from pointing straight down. The best way to determine true north is to take a bearing on a shadow at precisely local noon, when the sun comes from due south. But even this is complicated. To determine exactly when local noon is, we need to know our longitude with high precision, and that is challenging because this close to the *geographic* pole, lines of longitude become crowded together. And we need a sunny day when we're not deep in a shaded valley at noon. On this day, everything has aligned, and we find that the magnetic declination, or difference between magnetic and geographic north, is 118° SW. We will apply that correction to the readings we take on geological features in this tectonically complex area and check in on the unreliable pole again, if possible, in a few days. We've had a quick glimpse at the intensely deformed rocks—deep sea sediments called "turbidites"—and if we are to make any sense of them, it's essential to get properly oriented.

I'm getting my personal bearings back too. After finishing my PhD, I felt as if I were standing, unstably, at the edge of a precipice. Wounded and ashamed by my failed marriage, I was also deeply unsure about whether I could survive in the ruthless world of academic research. I yearned to continue working in the Arctic, but now that my doctoral project had ended, I was on my own and would somehow have to fund my own expeditions. Geologists who could simply get in a car and drive to their field sites had a great advantage. In the early years of an academic career, research productivity, measured by the number of papers published in prominent journals, is essential. I had enough data from my PhD work for several more papers, but in the long term I would need to build a new research program and stake a claim to some well-defined intellectual territory.

Among the job listings in the back of a professional newsletter, I spotted an announcement for a new post-doctoral position at the Ohio State University's Institute for Polar Studies, endowed by the family of aviator and Antarctic explorer Admiral Richard E. Byrd. In my application, I proposed to develop a numerical model to simulate certain deformational features I had observed in the Kapp Lyell diamictites in Svalbard—some distinctive ripple-like folds adjacent to dropstone clasts that seemed to have been caused by the mechanical contrast between the stones and the fine matrix. Although this was a rather narrowly focused project that had no great implications for polar science, it did involve ice-related rocks in a high-latitude place—and to my good fortune, the search committee named me the first Byrd Post-doctoral Research Fellow.

The fellowship provided a reasonable living stipend and modest research budget. I had my own office adjacent to that of ice-coring pioneer Lonnie Thompson, whose team was teasing out the records of climate fluctuation encoded in glacial ice from the Andes. Conversations with members of that group about how sensitively past temperatures had tracked greenhouse gas levels were my first real exposure to the looming fact of anthropogenic climate change. There were weekly brown-bag seminars on glaciology, polar ecosystems, and Antarctic geology, and the atmosphere was amiable. But most people affiliated with the Institute worked in the Antarctic rather than the Arctic, and as neither a student nor a faculty member, I had no formal framework for regular interaction with anyone. As a result, I spent much of my time alone and immersed myself obsessively in my project.

In my field of structural geology, which concerns rock behavior more than a million-year timescales, solid rocks in the middle and lower crust can be considered viscous fluids, and deformation of rocks during mountain building can be simulated using the mathematics of fluid mechanics. The motion of a viscous fluid—the "flow field"—can be described in two ways, both named for eminent eighteenth-century mathematicians: Joseph-Louis Lagrange and Leonhard Euler. In the Lagrangian approach, the velocities and paths of individual particles are tracked through time, while in the Eulerian method, indistinguishable particles flow into and out of a space in which velocities are specified at every point. Both are valid descriptions of complex flows, but they yield fundamentally different insights.

Working alone on a computer program to re-create the structures in my diamictites, I ruminated about how the Lagrangian and Eulerian approaches embody the tension between individuals and society and map onto underlying differences in modern versus traditional cultures. The Lagrangian frame, focusing on the path of the individual, has been the predominant worldview in Western culture since the Enlightenment. It has shaped Western literature: John Updike famously defined the novel as an "individual moral adventure"[1]—the solitary hero traversing rugged psychological terrain. The Eulerian perspective emphasizes patterns across generations, how all people pass through the same straits and eddies in a never-ending stream. This is the worldview of premodern societies, and the realm of fable and myth.

The Eulerian approach was more appropriate for the scale of rock features I was trying to model—and was also what I desperately needed in my own personal life. I longed for relief from my solitary Lagrangian existence. When I couldn't stand another moment alone with my inelegant computer code, I would go to the reading room in the Institute's library and try to learn more about how my work in Svalbard fit into the larger puzzle of high Arctic tectonics.

One day I came across recently published reports from the Geological Survey of Canada on the rocks of Ellesmere Island and was struck by similarities between the stratigraphic sequence found on the very northern edge of Ellesmere and the rocks, including diamictites, I had studied in Svalbard. The author of the reports, Hans Trettin, speculated that the distinctive rocks of northernmost Ellesmere were an "accreted terrane"—a tectonic addition that had come from elsewhere, like a modular home delivered, fully constructed, to a new site. ("Terrane," not "terrain," is the spelling used to denote such a region). He named this imported landmass "Pearya" for the (now-discredited) polar explorer Robert Peary. I was quite sure I knew where Pearya had come from.

I sent a short, timid letter to Trettin at his office in Calgary, describing my work in Svalbard and offering to share information about the rock units there. It felt like tossing a bottle with a message into the ocean; I didn't expect a reply. Within a week, Trettin had written back inviting me to join an expedition that coming summer in northern Ellesmere. He suggested that I come to Calgary in the spring—after the city was finished hosting the winter Olympics—to begin preparations for the field season. Receiving that letter was my lottery-winning moment, an inflection point in my life. I will always be grateful to Hans Trettin for bringing me to beautiful Ellesmere, at the far edge of the world. I would spend two summers as part of his Geological Survey mapping team.

Ellesmere is the world's tenth largest island, edged out for ninth place by Great Britain. Its northernmost promontory, Cape Columbia, latitude 83°06' N, is one of the most northerly points of land on Earth. In the Inuktitut language, it is *Umingmak Nuna*, land of musk oxen, and even now those formidable beasts outnumber people.[2] Remnants of whalebone houses as far north as 81° are thought to be seasonal camps built by people from the trans-Arctic "Thule culture" at about 1000 CE, during the Medieval Warm Period—the same time Vikings were sailing around the North Atlantic region. These settlements were abandoned as the climate cooled, and human presence on Ellesmere remained minimal until the twentieth century. In 1953, the Canadian government created the Inuit town of Grise Fjord on the island's southernmost point to achieve two political goals: affirming the country's jurisdiction over the island and forcefully relocating Indigenous nomadic people of the far north into permanent settlements. Some families were moved from the relatively hospitable landscapes in northern Quebec, with an abundance of berries, birds, and shellfish, to the rocky barrens of Ellesmere. Two other "communities," Eureka and Alert, were established early in the Cold War era as military and meteorological bases.

The Canadian Arctic Archipelago is vast, and starting in the 1960s, the Geological Survey had assigned huge tracts of it to a handful of geologists to map at a reconnaissance scale.

Trettin's domain included all of northernmost Ellesmere—
ten thousand square miles, a region larger than the state of
Vermont, without any human infrastructure. He had been
mapping largely by himself for years and, nearing retirement,
was hoping to understand how the deformational history of
northernmost Ellesmere fit into the larger tectonic puzzle of
the Arctic region. My letter had arrived at just the right time.
As it happened, Yoshihide Ohta—Yoda—of the Norwegian
Polar Institute had also been intrigued by the similarities between the rocks of Northern Ellesmere and western Svalbard
and had separately contacted Trettin. So two years after meeting Ohta in Oslo, I found myself stranded with him at the
Canadian Polar Continental Shelf Project logistical station in
the tiny town of Resolute on Cornwallis Island, hoping that a
nasty weather system would soon clear so we could be flown
to Trettin's base camp at Tanquary Fjord on Ellesmere five
hundred miles to the north.

There is nothing to be done in bad Arctic weather but wait
for it to get less bad. We tried to relax and savor the empty
time, aware that this would be our last chance in two months
to take showers, sleep in real beds, and be connected to the
outside world. But after thirty-six hours in the barracks-like
station with a score of other scientists, claustrophobia drove us
outside. Resolute has only a few streets, and it didn't take long
to explore the entire town. Polar bear pelts were draped on
porch railings outside many of the prefab houses, with the
black, heat-absorbing skin visible under the white fur. Although there were only a few remnant patches of snow from
the previous winter, snowmobiles were parked haphazardly
everywhere.

From a high point at the edge of town, we could see

Beechey Island, where archaeological teams had recently discovered the remains of three members of the Franklin expedition, which set off from London in 1845 in search of
the Northwest Passage and never returned.[3] The bodies had
been well preserved in the permafrost, and autopsies suggested these hapless men had died of lead poisoning from
their tinned food but were also weakened by scurvy, starvation, and hypothermia. I stopped at the co-op grocery store
to get an orange, a chocolate bar, and a spare "toque," a wool
cap. A five- or six-year-old Inuit girl joined me in eyeing up
the candy options. I asked her name, and she said "Gloria,"
which made me suddenly homesick because that is my mother's name. Her father came into the store and was curious
about where I was from. When I told him I had grown up in
Wisconsin, he replied, "Oh, WAY down south." It was a bit of
a shock to realize that was accurate. We were thirty degrees of
latitude north of my hometown; an equivalent distance to the
south would be somewhere in Honduras.

By the end of the third day in Resolute, I had already finished two of the six books I had brought with me for the
entire summer and was getting worried that I would run out
of things to read. The weather was improving, but other field
parties—oceanographers, biologists, and archaeologists as
well as geologists—were to be flown out before our group.
The first to go was an interdisciplinary team destined for the
"Ice Island" station off the northwest coast of Ellesmere, a
scientific base on a drifting raft of sea ice a hundred feet thick
and nine square miles in area. Finally, on the fourth day, it was
our turn. We loaded all of our gear into a gutted Douglas DC-
3—the classic plane that had launched passenger air travel in
the 1930s—strapping everything securely to the interior walls

of the cabin. That year, Tanquary was becoming the headquarters for the new Quttinirpaaq National Park, and the cargo also included furniture and materials for the park buildings. We humans, a dozen in all, found places on the floor amid the crates and lumber where we could sit during the cold four-hour flight.

It was too noisy in the uninsulated plane to talk with one another, so we could only point at the exquisite geometries in the landscape below—zigzags created by folded bedrock; lobate glaciers spreading over the terrain like thick pancake batter; pockmarked sea ice resembling the spongy part of bone. There was nothing in the landscape to indicate scale, and the mental effort of observing and interpreting so many new and unfamiliar features was strangely exhausting, like struggling to read a text in a language one has just begun to learn.

Finally we could see Tanquary Fjord, once a remote Canadian Air Force camp, with an airstrip that was simply the floodplain of a glacial river rushing out of the mountains. As we descended, the distance between the edge of the water and the base of the peaks looked alarmingly short, and the ground altogether too rough, but our pilot landed the plane as if we were on a routine commercial flight.

Two full days of logistics followed: setting up camp, organizing supplies, prioritizing the geologic work to be done. Unlike Svalbard, where food stores had to be protected from polar bears, the primary mammalian menace on Ellesmere were ravenous lemmings, and all provisions were kept sealed in sturdy boxes and drums. Water for drinking and cooking came from a small pond created by a weir across a spring-fed stream. The camp was far better equipped than any I'd known in Svalbard. Large Quonset hut–like tents on wooden platforms remained from the air force days, with ceilings high enough to stand up in without crouching. There was a cook tent—and a cook!—as well as a work tent with tables and actual chairs, where maps and air photos could be comfortably studied and annotated.

Tanquary brought together an unlikely collection of people, with at least six languages spoken among the dozen or so summer residents. Trettin was a native German speaker, and his enunciation, combined with Ohta's Japanese-inflected English, sometimes led to mutual incomprehension. Physically, the two were almost comical opposites: Ohta short, round, and mostly bald; Hans tall and lanky, with a shock of white hair. Ohta was also a gastronome while Trettin was an ascetic. What they shared was a profound connection to rocks that was as intuitive as it was scientific.

In addition to Trettin's group, two other Survey geologists—a French-speaking stratigrapher and a paleontologist hosting two Russians—would use Tanquary as their base for the summer. Three undergraduate students studying geology at Canadian universities served as field assistants. A fourth had been hired as camp manager, keeping track of supplies, managing the water supply, and coordinating the helicopter schedule for the various scientific subgroups. Inuit workers constructing park buildings had their own quarters but joined us under the midnight sun to play baseball, with two-by-fours for bats and a rock stuffed into wadded socks for a ball. The helicopter team, an unflappable Vietnam vet and younger copilot, arrived after flying an incredible two thousand miles from southern Alberta, hopping from fuel cache to fuel cache. At last, after months of preparation, and more than a week since leaving Calgary, it was time to get to work.

The camp at Tanquary was more than sixty miles from our areas of geologic focus on the north coast of Ellesmere, so each lead geologist would be flown by helicopter for a week or ten days with a field assistant to remote "fly camps," then return to the base to rest and enjoy the luxuriant amenities there. Hans, in what was probably a violation of Geological Survey safety rules, preferred to work alone at remote camps, covering vast stretches of territory with his long legs and living off little more than oatmeal.

At fly camp, we would stay in heavy, pyramidal "Scott" tents—the type used by the ill-fated Robert Falcon Scott in his quest to reach the South Pole. This type of tent has a wide skirt around the base, which we would anchor with rocks in anticipation of high winds. The tents were all one piece, with their tall poles sewn permanently into the heavy canvas fabric. They were much too long to fit inside the helicopter and had to be strapped securely to the landing skids. At first, I disliked these unwieldly tents; they admitted almost no light, and it seemed absurd to have to use a lantern in a place where the sun wouldn't be setting for two months. Also, the entrance was a floppy tube of canvas, like the top of a sock, and crawling in and out, especially with a backpack on, was awkward and undignified. But after our first storm with sixty-knot gusts, I came to appreciate the wind-worthiness of the design. I remembered, all too vividly, lying in my tiny tent in Svalbard in a gale, with the fabric pressed against my face and the poles bowed to the point of snapping. By comparison, our Scott tent seemed like a brick house.

That first summer on Ellesmere, Trettin asked me to map the strongly deformed "suture" zone between northernmost Canada and the Pearya terrane—the ancient plate boundary where a continental fragment that may have once been part of Svalbard was accreted to North America during the formation of Pangaea. Trettin had shown that the rocks of Pearya had a geologic history completely different from the native North American ones to the south. But the area in between those well-defined crustal blocks, along Kulutingwak Fjord, had not been mapped in detail. So on a bright, brisk day, a field assistant and I were dropped off by helicopter into the middle of a tectonic mess at the top of the world. The first step was to get our magnetic bearings, and then we could start trying to make sense of the contorted rocks. We soon discovered that figuring out which way was true north was simple compared with determining which way was "up" in

Unnamed valley along Kulutingwak Fjord, northern Ellesmere Island.
The mountains are made entirely of turbidites.

the strongly folded and faulted sedimentary sequence, which consisted almost entirely of a single rock type: turbidites.

Turbidites are distinctive, repetitively layered, sedimentary rocks formed when unstable piles of sediment at the edge of a continental shelf unleash dense, fast-moving slurries of sand and mud—called "turbidity currents" or "flows"—that hurtle down submarine canyons and out onto the deep sea-floor. Turbidity currents are the only mechanism by which significant volumes of continentally derived sediment reach the deep ocean. A single turbidite bed may be a few inches to several feet thick, but turbidites are gregarious rocks, tending to occur in droves. Wherever they are found, they form sequences hundreds to thousands of feet thick, documenting the passage of many submarine flows at a given place over time. As turbidity currents lose velocity, they leave behind progressively finer sediments, creating a characteristic pattern called "graded" bedding. In places like northern Ellesmere, where such sediments have later become tilted, deformed, and even inverted, these size-sorted layers may be the only means to decipher the "younging" or original upward direction of the strata. Finding clear instances of graded bedding within a heap of folded and faulted rocks is like being hopelessly lost in a dense forest and then spotting a well-marked trail.

Turbidites hold a special place in the intellectual history of geology, and in fact appear at the site that is considered the birthplace of the modern discipline: Siccar Point, Scotland. This wave-battered promontory hosts one of the most famous rock outcrops in the world, the site where, in 1788, the Enlightenment Era polymath James Hutton is said to have discovered "deep time." At Siccar Point, strongly folded, steeply tilted turbidites are truncated by an irregular erosional surface—now called an "unconformity"—and overlain by a younger, planar sequence of strata. Hutton called the tilted turbidites "the ancient schistus" and had no inkling of their deposition by powerful submarine flows. But he correctly interpreted the rocks' deformed state as evidence that they had once been deep in the heart of an ancient mountain range. A century and a half before geologists developed the concept of plate tectonics, this was an astonishingly insightful inference. Hutton further surmised that the discontinuity between the tilted and undeformed rocks—the unconformity surface—represented the time it would take for erosion to wear that mountain range to a flat plain. The unconformity at Siccar Point revealed planetary timescales that were much vaster than Western thinkers had previously imagined, and it became the foundation on which the modern science of geology was built. Yet many decades would pass before geologists discovered the processes that had laid down the turbidites below the unconformity.

The story of the evolving interpretation of turbidites is used as a kind of moral lesson for geology students—a cautionary tale about the dangers of naive overinterpretation. By the mid-nineteenth century, the rocks now called turbidites had been mapped and described in many mountain belts around the world, notably the Alps, Appalachians, and British Caledonides. Geologists working in the Swiss Alps in the early nineteenth century seem to have been the first to denote turbidites as a distinct rock type. They called them *flysch*, a Swiss German term that may come from the word *Fleisch*, meaning meat or flesh, possibly an allusion to the striped pattern created by the repeating beds. In places, Alpine flysch sequences include jumbled intervals with large blocks of other rock types, and these were called "wildflysch," aptly suggesting something feral in their nature (today, these are interpreted as undersea landslide deposits). German geologists, meanwhile, referred to turbidites as *grauwacke*, meaning simply "gray stone." The rather somber tone of this name was lost when it was imported into English as "greywacke," with the second half pronounced "wacky."

Early geologists were puzzled at the observation that flysch/greywacke sequences were found only in mountain belts—invariably in a folded and faulted state—and never as undisturbed horizontal beds. They did understand that the immense thickness of these sequences meant that they had been laid down in deep basins that could accommodate large amounts of sediment. But Charles Lyell's doctrine of uniformitarianism—looking to present-day processes to explain how rocks formed in the past—seemed to fail in the case of flysch. There was no observed phenomenon that could deposit such accumulations of repetitively layered rock, and no known mechanism to elevate them high into mountain belts.

Turbidites became a central component in an elaborate, misguided paradigm for mountain building: the theory of "geosynclines." Because flysch was found in all major folded mountain belts but did not occur in other settings, geologists surmised that there must be a causal connection—namely, that flysch was the agent of mountain building. Although there were different versions of geosynclinal theory over the years, the essential idea was that as sediment accumulated in a deep submarine trough called a geosyncline, it would eventually reach critical thickness, became unstable, and . . . somehow give rise to mountains. Incredibly, this was the prevailing conceptual model for mountain formation right up to the emergence of plate tectonics in the 1960s.[4]

The mystery of the depositional origin of flysch had been solved a few decades earlier. The critical clues arrived unexpectedly in 1929, when a major earthquake occurred off the south coast of Newfoundland, far beneath the waters of the Grand Banks cod fishery. Less than a minute after the earthquake, twelve telegraph cables were cut in rapid succession by something powerful that swept across the seafloor at up to sixty miles per hour.[5] Two hours later, a series of tsunami waves devastated fishing villages in the narrow inlets of Newfoundland's Burin Peninsula, killing twenty-eight people. It was later recognized that the earthquake had triggered an immense submarine landslide, which not only generated the tsunami but also let loose the leviathan that had cut the telegraph cables—a roiling juggernaut of sediment cascading off the continental shelf onto the abyssal plane of the seafloor: a turbidity current.

After being interrupted by the wars, the study of turbidites and turbidity currents became one of the hottest areas of geological research in the 1950s. The discovery of turbidity currents did not directly challenge geosynclinal theory, but the study of them prepared the field of geology in two important ways for the plate tectonic revolution. First, it helped reunify a science that had become splintered into insular subcultures with separate research agendas. Sedimentologists studying ancient turbidites in the rock record began to correspond with seismologists and oceanographers, who could provide information about modern processes in the deep sea. Hydrologists who had done scale modeling of river systems adapted their laboratory flumes for turbidity current simulations.

Second, the discovery of turbidity currents was a sobering reminder that geologic processes in the deep oceanic realm could be fundamentally different from those observable on land or even in the shallow waters of the continental shelf. Geologists had to acknowledge that decades of careful observance of Lyellian uniformitarianism might have led to a blinkered view of the full range of Earth's habits; the planet's behavior might be wilder and stranger than previously thought. This realization may have primed geologists subliminally for the plate tectonics revolution, which required visualizing processes in the deep ocean that had not been observed directly: seafloor spreading and subduction.

Subduction—or more specifically, collisions between continents that occur once the oceanic crust between them is subducted—was the long-sought mechanism for building mountains, and within a few years of its discovery, almost all geologists had abandoned the increasingly untenable concept of geosynclines. The reason that turbidites were exclusively found in mountain belts was that mountains were the only place land-based geologists would ever encounter them. And their presence in mountain belts around the world simply reflected the fact that they happened to be on the front lines when continents plowed into each other. In 1970, Peter Coney, a structural geologist at the University of Arizona, commented wryly: "Saying geosynclines lead to orogeny [mountain-building] is like saying fenders lead to automobile accidents."

But even after the plate tectonics paradigm was adopted, it took some time for the place of turbidites in the tectonic system to come into focus. By the mid-1970s, deep-sea drilling and geophysical imaging of subduction zones had revealed a distinct tectonic habitat for turbidites: the "accretionary wedge" or "prism."[6] As a subducting ocean plate slides beneath an over-riding one, the turbiditic sediments carried by the down-going plate are often scraped off and accreted onto the base of the upper plate. As this process continues, a thickening wedge of sediment builds up at the interface between the plates, like the triangular prism of snow in front of a plow, with the most re-cently added material at the leading edge.

Accretionary prisms occur at many modern subduction zones, including those off the coasts of Japan, the Philippines, Indonesia, and the "Cascadia" region of the Pacific Northwest in the United States. In all these places, the watery, fault-riddled wedges of sediment—and their frictional contact with the sinking plate—determine the magnitude and frequency of great subduction-zone earthquakes. The island of Taiwan is also part of an active accretionary prism above a subduction zone, a rare instance in which the wedge of scraped-off sedi-ment has become thick enough to emerge above sea level. Ancient accretionary prisms can form larger landmasses; most of New Zealand's South Island is composed of Mesozoic-age turbidites that were amalgamated into a large accretionary wedge complex during a long period of subduction between 140 and 100 million years ago, entirely distinct from New Zealand's present tectonic setting.

The turbidites we were studying on Ellesmere Island were probably part of an accretionary prism formed 400 million years ago when subduction brought the Pearya terrane—or a larger continent to which it was attached—inexorably closer to the edge of ancient North America. At the time of our work in Ellesmere Island, the interior architecture of modern accretionary prisms was just beginning to be explored through seismic surveys and drilling projects in the Cascadia prism in deep water off the coast of Oregon, and it was fascinating to compare indirect observations from that active but inaccessible wedge with the features we could see in the ancient turbidites beautifully exposed in the mountains around Kulutingwak Fjord.

Most notably, many of the structures in the turbidites on northern Ellesmere appear to have formed before the sediments had been fully compacted and cemented into solid rock. These "prelithification" structures included dikes—or sheetlike intrusions—of sand that cut across the layering in the rocks. Normally, dikes are igneous features, formed when magma is forcefully injected into cracks in preexisting rocks. In contrast, these sand-filled dikes were made of the same material as the surrounding turbidites.

Fortunately, I had seen such "clastic dikes" before, in two-billion-year-old turbidites on a graduate school field trip to the Upper Peninsula of Michigan. Geology professors often cajole students into action with the adage that "the best geologist is the one who has seen the most rocks." While I dislike its competitive undertone, I can't deny the truth of the maxim: the more rocks we get to know, the more likely we are to understand ones we'll meet later. When I saw the sand dikes in Ellesmere, I knew what they were and how they had formed: turbidity currents had piled up sediment so rapidly that it didn't have a chance to compact and lose the water trapped between the grains. At some point, water-saturated sand at the bottom of the pile became so overpressurized that it shot up violently through fissures in the overlying sediments. A dramatic illustration of this phenomenon is in evidence today at the notorious Lusi mud "volcano" in East Java, Indonesia, where water-laden clay from a source deep underground has been spewing from a fissure since 2006, drowning villages and forcing the evacuation of forty thousand people. On Ellesmere, we could sense the same fearsome power in those strange bodies of intrusive sand.

In the Ellesmere turbidites, the beds were commonly contorted into swirling patterns that are typical in rocks that have experienced ductile deformation at high temperatures—yet these rocks were barely metamorphosed. The most reasonable explanation for the convoluted layering was that it reflected disruption of partly cohesive but still-watery sediment as it was being incorporated into the accretionary prism. Core samples taken around the same time from the Cascadia prism contained such "soft-sediment'" structures at the scale of inches; in Kulutingwak Fjord, we could see them on mountainsides. We soon realized that we needed to be careful to distinguish these prelithification structures from later, more common types of deformational features—faults, veins, folds—that formed during the eventual collision between Pearya and North America, once the turbidites had become proper rocks and were being pushed up into mountains above sea level.

Little by little, we gained a feeling for the idiom of these rocks and how to disentangle the multiple narratives they contained. Finding clear graded bedding was always cause for jubilation because it was critical in defining the larger-scale geometry of the rocks, a difficult task in a monotonous sedimentary sequence without compositionally distinctive layers. The sheer volume of turbiditic sediment was astonishing—steeply tilted beds forming a chain of mountains seventy-five miles long and fifteen miles wide. It was hard to know what their original thickness and extent had been, however, because they had been telescoped into this gargantuan stack like tiles on a roof. Turbidites upon turbidites upon turbidites. We joked about feeling like we were in that Monty Python sketch about déjà-déjà-déjà vu until it started to feel too uncomfortably close to the truth.

Just when we were beginning to feel dispirited by the prospect of spending the rest of the summer trying to map an undifferentiable mass of turbidites, we encountered two weird and wonderful sites with completely different rocks. One was a jumbled zone with angular, boulder-size fragments of volcanic rock—what the old Swiss geologists would have called *wildflysch*. It was enough to send us into an ecstatic wild rumpus. (Unfortunately, in my view, the charming old Swiss term for this kind of coarse, disorganized material has been abandoned in favor of a tongue-tangling Greek-derived neologism: olistostrome, or "slide bed"). The occurrence of this isolated bouldery interval within the turbidites suggested that an unusual incident—perhaps a great earthquake—had caused an avalanche of rock to be shed into deep water from the flanks of a nearby subduction-related volcanic island, similar to modern-day Sumatra or Honshu. This implied, in turn, that the turbidites had been deposited close to the subduction zone that had stacked them into an accretionary prism, which would account for the deformation they had experienced while still in a watery state. Even more important, the volcanic boulders contained minerals that could be isotopically dated, and could thereby constrain the moment in geologic time when all these events had transpired. We filled our backpacks with samples for Hans to submit to the Geological Survey's geochronology lab.

An even more marvelous discovery was a sea-green knoll

that we spotted from a distance after climbing up over a high pass into a broad, undulating valley. We kept peering with binoculars at the strange hill, trying to guess what could possibly make it green. Vegetation was so sparse at this latitude that we ruled that out immediately. Instead, it had to be rocks that weren't turbidites, and that alluring thought drew us on. When we reached the little rise, we were astonished to find that it was made entirely of serpentinite—an exotic rock type that has its origins in Earth's mantle and whose slippery, scaly surface is a bit reptilian. The discovery was so completely unexpected and ridiculous that at first we simply laughed. This rock was far out of place. Stumbling upon it amid all those sandy turbidites was akin to encountering a giant squid in the Sahara, or maybe a desert snake on the deep seafloor.

Serpentinite is a hydrated form of the igneous rock called peridotite, named for its primary mineral, olive-green olivine, whose gem form is peridot. As the rock type that makes up much of the Earth's mantle, peridotite is the planet's most abundant rock by volume on Earth, but it takes very unusual tectonic circumstances for it to reach the surface. Because its natural habitat is Earth's hot interior, peridotite is unstable in the presence of water, and will react with it to form serpentinite. Most exposures of peridotite (and serpentinite) occur in rare places around the globe, where subduction somehow went awry and slabs of oceanic lithosphere were thrust onto continental crust. These "misshelved" rock complexes are called ophiolites, and famous examples occur in Newfoundland, Northern California, Oman, and Cyprus. But our accretionary prism complex at Kulutingwak Fjord was not an ophiolite. Here, somehow, serpentinized mantle material had found its way into a pile of turbiditic seafloor sediments.

I'd seen serpentinites in California's Coast Range ophiolite during my internship with the US Geological Survey and recalled learning then that serpentinite has the unusual ability to flow in the solid state, in a manner akin to glacial ice or rock salt. Because its density is lower than the peridotite from which it forms, it will tend to rise upward, snaking its way through cracks and fissures in the crust and sometimes erupt in ultra-slow motion as a cold mass, at Earth's surface or onto the seafloor. We could tell that our serpentinite had pushed through the turbiditic sediments while they were still accumulating on the deep ocean floor because there were pebbles and cobbles of serpentinite incorporated into adjacent turbidite layers. This was clear evidence that the serpentinite had been exposed as a mound on the seabed and that chunks had rolled down its flanks to be incorporated into the sediments.

Around the time we encountered this bizarre blob of alien rock, a scientific drilling project had just documented submarine mountains of serpentinite hundreds of feet high in the area west of the Mariana Trench, a subduction zone in the Pacific Ocean—and the deepest point on Earth's solid surface, at six miles below sea level. The interpretation of these seamounts was that water carried by the subducting slab had hydrated the overlying mantle rock and changed it to serpentinite, which then found passageways up through the ocean crust and out onto the seabed.[7]

On Ellesmere, we could examine such a seamount in person without a submersible. We spent several hours marveling at the knob of serpentinite, inspecting it from all angles, until late in the day we reluctantly bid it goodbye. As we made the long trek back to camp, I realized that my conception of the monotonous turbidites had shifted. I had a vision of the seafloor where they accumulated as a tumultuous marketplace where goods from disparate sources were traded. From the lands above sea level came far-traveled continental sediments as well as rough, raw materials contributed by nearby volcanic islands. From the subterranean realms came overpressured sands and mantle-derived serpentinite. These rocks were an open system full of energy and motion.

We documented the wild olistostrome and serendipitous serpentinite with photographs and sketches and collected dozens of samples in official Geological Survey of Canada cloth bags with labels reading ON HER MAJESTY'S SERVICE, as if we were part of some sort of high-stakes spy mission. But spying was not the right metaphor for what we were doing; instead we were students—schoolchildren, really—learning to read a text written at the scale of the landscape, about a vanished world that was both unreachable and vividly present.

Sometimes, when our heads became too heavy with rocks, we needed some frivolity. One of the field assistants, a Quebecker, would help me practice my French by reading trivia questions printed on packets of instant oatmeal. Most of them had to do with hockey, and I found that even if I didn't catch all the words, I'd be right about half the time by answering "Wayne Gretsky." We had a nightly radio check-in with the crew at Tanquary and other flycamps. Sometimes the signal was bad and we would have to spell out essential messages letter by letter using the NATO radio alphabet (Alpha, Bravo, Charlie, Delta, Echo, etc). When reception was better, there was an ongoing competition among the various field parties to create the most confusing possible version of the alphabet, with bonus points for incorporating geologic terms. For example, the code words for the *G*, *K*, and *N* were Gnu, Knew, and New; *M* was "mnemonic" and *T* was "tsunami." There were several nominees for *P*, including "psychrometer" and "pterodactyl," but the undisputed winner was "pneumonoultramicroscopicsilicovolcanoconiosis," an inflammatory lung disease that results from breathing volcanic ash.

After spending so much time in the company of turbidites on Ellesmere Island, whenever I have encountered them in other parts of the world—Scotland, Italy, New Zealand, northern Wisconsin—they seem like old friends. I understand how to read them; I'm always glad to hear their familiar stories. But in a Zen-like paradox, the turbidites that are no longer with us—the ones that do get subducted—actually play an even more fundamental role in the Earth system. As the only continentally derived materials to reach the deep-sea floor, turbidites are the primary means by which continental crust can be recycled.

Earth's basaltic oceanic crust is disposed of quite easily: Born at midocean ridges by decompression melting of the mantle, it returns to its origins via subduction about 150 million years later when it is older, colder, and denser. Continental crust, however, no matter how old, is too buoyant to be subducted. The continents are more heterogeneous in makeup than the uniformly basaltic ocean crust, but to a first approximation, the continents are granitic, and granite is far too light to sink into the dense peridotite of the mantle. Even erosion can't completely get rid of continental crust because most eroded sediments end up on the continental shelves, which, though below sea level, are still underlain by continental crust.

Turbidity currents are the one way that continental material can reach the abyssal ocean floor and have a chance of being subducted. While much of this material ends up in accretionary prisms—and, eventually, mountain belts back on land, to be puzzled over by geologists—perhaps half of it gets subducted along with the down-going ocean plate.[8] In this way, turbidites "close the loop" for continental crust. Amazingly, this rather Rube Goldbergian system has kept pace with the rate of creation of new continental crust through subduction-related volcanism, and the total volume of the continents has remained about constant for almost 3 billion years.[9] If I were to make a list of Earth's most remarkable attributes, the role of turbidites in recycling continental crust would be close to the top. In turbidites, so long misunderstood, we have a glimpse of the subtle rocky logic of the Earth.

Those summers on Ellesmere, living as much on the deep-sea floor as along an Arctic fjord, expanded my geologic imagination and taught me to trust my scientific instincts even when I thought I might be lost. Convoluted turbidites in a place that was north of north helped me get oriented again after a period when I couldn't see the path ahead. Sometimes there is nothing to do but rely on one's inner compass and keep recording observations in the hope that eventually it will all make sense.

6

DOLOMITE

On a leaden winter day in southwestern Ohio, I'm standing on the sloping side of a reservoir, trying to show interest in gray rock fragments left from the construction of the dam that created this artificial lake. A faculty colleague has generously brought me here to share possible field trip destinations for the classes I will be teaching. The rocks are dolomite, or more properly dolostone, a cousin to limestone. They're full of fossils: brachiopods ("lamp shells"), broken stems of crinoids (stalked organisms related to starfish), bryozoa (branching, coral-like creatures), as well as everybody's favorite, trilobites. It's a glimpse of everyday life on the seafloor 450 million years ago. These Cincinnatian Series strata are famous in the world of paleontology. Their fossils serve as a kind of reference set for the late Ordovician Period. I respect these dolomitic rocks and am grateful for all the carbon dioxide they store in mineral form, but for some reason I don't feel any affinity for them. It's not their fault, but even the name dolomite feels dreary, evoking the doldrums, the Via Dolorosa.

In the months before I first went to Ellesmere Island, I'd submitted applications for the few academic job openings in structural geology that year. Many of my graduate school friends had taken lucrative jobs in oil and gas exploration and were making two to three times the salaries being offered for tenure-track assistant professors. They urged me to consider joining them in Houston, if only for a year or two.

Geologists, both individually and collectively, have a complicated relationship with the petroleum and mineral industries. Many people are attracted to geology as students out of love for fieldwork in wild places and then face an ethical dilemma when they find that the best-paying jobs contribute to environmental destruction. From its start, the discipline of geology has benefitted monetarily and scientifically from its alignment with the business of resource extraction. Petroleum and mining companies have provided scholarships for students and research grants to faculty. It's an uncomfortable fact that much of what we know about sedimentary rocks, paleontology, evolutionary patterns, and even long-term climate change has emerged from decades of research by oil companies. One of my friends compared leaving the frugal realm of academic research for a position in Big Oil, with its rich datasets and extraordinary computing resources, to the moment in *The Wizard of Oz* when the world changes from black-and-white to Technicolor. I struggle with the terrible irony that our worst environmental practices have also led to some of our deepest insights about how the planet works. Even when we treat the earth with disrespect, it replies with wisdom.

Personally, I couldn't imagine myself in the corporate world, or in a city where an inch or two of snow was considered a calamity. Also, fresh academic credentials have a short shelf life, and if one doesn't land a faculty position within a few years of receiving a PhD, a university career becomes less and less likely. As it happened, all the universities I applied to invited me for job interviews, and I was lucky enough to get three good offers. None of these were in places I had ever imagined I might live, but of the options, Oxford, Ohio, with its aspirational name and ivy-covered red brick buildings, seemed least unimaginable. I liked the fact that the university was home to the progressive Western College for Women, where civil rights activists received training before heading to Mississippi for the Freedom Summer of 1964. Also, the landscape was interesting; the deep valleys incised by tributaries to the Ohio River promised good bike riding—and there was some possibility of skiable snow in the winter.

I chose to look past some early troubling signs, like the fact that the university's mascot used an offensive term for Native Americans, as if native people existed only in the realm of myth. Allying myself with that culture felt like a betrayal of my sister, whose life was in tumult. She had recently fled an abusive relationship for somewhere in California and left her three small children with an acquaintance. For more than a year we did not know exactly where she had gone. Wherever she was, she was in pain. The anguishing reality of her experiences juxtaposed with the cartoonish mascot logo made me queasy.

My whole interview for the position had begun disconcertingly, with the faculty member who was to take me to breakfast bringing me to his house, opening the refrigerator, and asking me what I wanted to eat. There were other awkward moments in the two-day interview, which I would later recognize as passive-aggressive psychological tests. Did I know of a paper one of the professors had recently published and what did I think of it? How would I manage multiday field trips with the undergraduate geology majors, who were primarily male? I must have provided acceptable answers, since they offered me the job. Because the Ellesmere field season lasted into early September of that year, I started the position midyear in the second semester. As a result, I missed new faculty orientation events and the opportunity to meet peers outside the geology department, which I look back on with regret.

At twenty-six, I was more than a decade younger than any of my male colleagues, and I felt highly exposed, constantly under their scrutiny. My first few weeks of teaching were so stressful that I considered quitting; I'd been lucky in graduate school to have a fellowship that exempted me from being a teaching assistant, but now my lack of classroom experience was a liability. I was only a few years older than most students in my classes—and younger than many of the graduate students. Fraternities were a powerful social force on the campus, and there was a certain insolent male glower that could knock me off my train of thought in the middle of a lecture. I felt lonelier and somehow farther from home than I'd ever felt in Ellesmere or Svalbard.

One day when I was in the department office to pick up my mail, a beloved emeritus professor who often came in to chat with the two female administrative assistants called over to me and said, "Hey, little girl, come here and sit on my lap." Mortified, but also afraid of causing offense, I shot a look across to the women, who just shrugged and shook their heads. Red-faced and confused, I mumbled that I had to get ready for class and rushed out of the room.

I became dimly aware of long-running divisions within the department, whose origins I never fully understood. Everyone just seemed resigned to the fact that some people refused to interact with others. Several fellow faculty members were single divorcés, and I was on high alert for predatory behaviors, well aware of the dangers of romantic entanglements. But one colleague, N, a charming, soft-spoken Scottish ex-pat from Edinburgh, seemed to see through my skittishness. With him, I felt relaxed enough to have real conversations not only about geology—he had worked in southwest Greenland on some of the most ancient rocks in the world—but also about books and the arts, and the challenges of being aliens stranded in southern Ohio. N lived near campus, had a baby grand piano, and invited me to come over to play it any time. He and I shared a deep interest in the history and epistemology of geology, which our colleagues considered irrelevant to the modern practice of the science. We were also temperamentally alike; both of us dreaded social events like faculty dinners and preferred quiet evenings spent reading at home. It seemed natural, almost inevitable, that the two departmental oddballs would find each other. A serious, well-established academic, he was also a counterpoint to my reckless first husband. The only hitch: he was more than twice my age and, in fact, a bit older than my parents. Within the year, despite cautionary words from friends and family, we got married at the county courthouse.

N and I saw ourselves as equal partners and intellectual peers, but the outside world did not. An old friend of mine pointed out the parallel of our marriage with that of Dorothea Brooke and Edward Casaubon in *Middlemarch*, one of my favorite books, a comparison I shrugged off with irritation. I also pushed aside the obvious irony that two scientists in a discipline that concerned, above all else, the power of Time to "devour all things"[1] would consider themselves exempt from its effects. People who didn't know us thought I was his daughter, or when they learned we were married, assumed I had been his student.

In the geology department, I had effectively aligned myself on one side of the old schisms, and I became aware this could be a liability when I was considered for tenure. A tenure evaluation typically occurs six or seven years after one's appointment and is a singular, all-or-nothing moment of reckoning. Either one is granted tenure—and with it the security of a position one could hold until retirement—or one is ushered out the door, which usually means out of academe entirely, since a failed tenure bid is poison on a faculty job application. I realized that my marriage to N would require me to prove beyond any doubt that I was an independent scholar worthy of earning tenure. So I worked at a manic pace.

I conquered my shyness in front of the classroom by over-preparing for every lecture in the various courses I taught, from introductory geology to a graduate class in rock mechanics. I offered to update some of the "intro geo" laboratory exercises—still stuck in the old "rocks in boxes" mode—and ended up writing an entirely new lab manual that put the emphasis on Earth systems and processes rather than on collections of specimens. Since I was the sole advocate for this approach, I took over supervision of all the graduate TAs who taught the labs.

Although it was difficult to attract graduate students to study structural geology in an area where flat-lying dolostone stretched as far as the eye could see, I managed to recruit three good candidates in my second year, and had them working on three completely different research projects, supported by two successful National Science Foundation grant proposals. One of these students was Chinese, and I picked him up from the Cincinnati airport when he first arrived in the United States. An hour later, when we reached campus and the graduate student housing, where he would be living, he asked when he would be able to meet his advisor, Dr. Bjornerud. With a shock, I realized that during the entire drive he had thought I was a graduate student or department assistant—and that throughout our (preinternet) correspondence he had assumed I was much older—and male. After recovering from that rocky start, we developed a good working relationship. Around the same time, Ohta invited me to come back to Svalbard for a summer of mapping with the Norwegian Polar Institute near the Polish station at Hornsund. The following year, one of my new graduate students accompanied me for another season of work with Hans Trettin on Ellesmere Island. My department chair warned me of the dangers of spending too much time in the field and not managing to write up research results, so I published nine papers in three years.

In any case, it would soon be impossible for me to be gone all summer in the high Arctic. Two years before my tenure evaluation, I got pregnant with my first son, O, who was ac-commodating enough to arrive over the winter holiday break in a relatively easy delivery. My department chair advised me not to take any time off because a gap in my productivity might indicate to the tenure committee that I was not com-mitted to my job. So little O came with me to work every day for the first six months of his life. I nursed him in my office and had a roster of undergraduate students who babysat him there when I had classes or committee meetings. During labs and outdoor fieldwork, O peered over my shoulder in a backpack, enchanting the students. My husband, who had

had polio as a child, walked with a cane, so it was difficult for him to carry the baby safely. Also, N had fallen and broken his arm a week after O was born, making it impossible for him to change diapers, and this—in addition to the unavoidable asymmetry in parental care that comes with breastfeeding—set up an unfortunate pattern that would persist over time.

N had two daughters from two previous marriages and didn't share with me the maelstrom of conflicted feelings that come with being a parent for the first time. I felt whiplashed between a ferocious mother-bear instinct to protect my cub and a deep mourning for the independence I had given up for the tamed, constrained life that motherhood required.

But there was some virtue in not spending every summer doing fieldwork. Like me, N loved rocky, windswept North Atlantic places. We rented cottages atop the craggy granites of Cape Breton, the red sandstones of Orkney, and the ancient gneisses on the Isle of Lewis. When N's daughters came along, one of them not much younger than me, people were visibly confused at our family relationships. Our best times together were rambling along desolate rocky shorelines, musing about complexly deformed rocks.

These trips helped us to sustain the illusion that we didn't really live in southern Ohio, with all its flat gray dolostone. My feelings toward the town and the bedrock beneath it were strangely entangled. I was grateful to the university for providing me with a good livelihood, but somehow I could never fully enter into its logic and embrace the place as home. For me, life there felt muffled, as if the true nature of the place was behind a padded door. The idyllic appearance of the campus seemed like a euphemism that hid darker truths about what really went on in departments and frat houses. I found the

dolomitic bedrock similarly closed off, even though I appreciated that dolostone, as a rock that stores carbon dioxide in solid
form over geologic timescales, has played an essential function
in keeping Earth habitable—the nontrivial job of keeping the
planet from becoming a Venus-like hothouse. But I've always
found dolostone, with its cryptic, recrystallized texture, to be
maddeningly incommunicative. It's a rock that mumbles.

I'm not the only one to be frustrated by dolostone. Despite its ubiquitous occurrence, geologists have long found
its existence hard to explain. Most sedimentary rocks can
be understood by looking to settings where similar deposits
are accumulating today for insights into how ancient strata
formed. This is classic Lyellian uniformitarianism, but dolostone defies that logic. Dolostones such as the fossil-rich
Cincinnatian Series were clearly laid down in shallow marine
environments, but there are no known sites on the modern
continental shelves where great beds of dolostone are currently forming. The process must occur clandestinely, in the
subsurface, by the transformation of limestone into dolostone
sometime after the sediments have been deposited and buried.

Limestone consists mostly of the mineral calcite, or
$CaCO_3$, while dolostone is made mainly of dolomite
$CaMg(CO_3)_2$—so the process of transubstantiation must involve the introduction of magnesium, presumably by pore
waters moving through the sediment. But exactly how and
when this happens is not completely understood—and at-
tempts to grow dolomite in laboratory settings under the
temperature and pressure conditions where it seems to have
formed in nature have failed. Geologists call this "the dolomite problem," and it is rather embarrassing, given the central
role dolostone plays in the global carbon cycle.

Although we tend to think of plants and soils as the primary players in the carbon cycle, this is true only in the short term; carbonate rocks are actually in charge over the long haul. The total mass of carbon stored in plants is substantial, about 500 gigatons (a gigaton is a billion tons). Soils hold an even more impressive 2,500 gigatons. But carbonate rocks—the majority of which are dolostones—contain 100,000,000 (one hundred million) gigatons, 99.9 percent of all the near-surface carbon on Earth.

Another essential distinction between carbon in plants and soils versus rocks is the "residence time" of that carbon—the typical time it will spend in a particular form or "reservoir." The average residence time of carbon in plants is about five years, which reflects the global mix of annual species like grasses versus longer-lived trees and shrubs. Carbon in soils has a residence time of decades to centuries, depending on climate and land-use patterns. Together, plants and soils—representing the processes of photosynthesis and decomposition—are key players in what biogeochemists call the "fast carbon cycle." This is akin to ephemeral communications like emails and tweets, constantly exchanged and briefly important, but quickly forgotten. In contrast, the residence time of carbon in carbonate rocks is tens of millions of years—analogous to the enduring literature that comes to define a culture. This "slow carbon cycle" is one of Earth's most remarkable attributes, a steadying system of global checks and balances that involves the atmosphere, hydrosphere, biosphere, and lithosphere.

The long-term cycling of carbon begins with the slow leak of carbon dioxide (CO_2) out of the interior of the Earth from volcanoes. This deep-Earth CO_2 combines with water vapor in the atmosphere, making natural rain slightly acidic and able to dissolve ions like calcium out of rocks exposed on land. Importantly, this "weathering" process is most intense when atmospheric carbon dioxide levels and temperatures are high, drawing more CO_2 out of the atmosphere and cooling Earth down. Conversely, less intense weathering in cold times allows more CO_2 to accumulate in the atmosphere, which causes temperatures to rise. This temperature dependence of rock weathering rates makes the slow carbon cycle a remarkable, stabilizing planetary thermostat over geological timescales.

In the next stage of the slow carbon cycle, calcium ions derived from rock weathering are carried in solution to the oceans, where they meet more dissolved carbon dioxide in the form of bicarbonate (HCO_3^-)—and the two species of charged particles bond to form the mineral calcite. Notably, much of this matchmaking is done biologically, by marine organisms ranging from algae and corals to clams and starfish, who armor themselves with calcite. When these organisms die, their calcite shells and exoskeleta rain to the seafloor to become limestone, locking atmospheric CO_2 away in mineral form.

And then, somehow . . . dolostone forms.

Dolostone is unusual among rock types because its discovery can be attributed to a specific person: Frenchman Déodat de Dolomieu (1750–1801), named for his native town, who was swept up in the political intrigues of the French Revolution and Napoleonic Wars and spent two years languishing in a Sicilian prison. Although his primary focus was volcanism, during his expeditions to the Tyrolean Alps he noted the widespread occurrence of a rock that resembled limestone but did not effervesce vigorously when acid was applied to it, as limestone does. (This "acid test" breaks the bonds between calcium and carbonate ions in calcite, allowing long-sequestered carbon dioxide to bubble back out into the atmosphere.) In his 1791 paper "On a Type of Calcareous Rock That Reacts Very Slightly with Acid,"[2] Dolomieu introduced the recalcitrant "new" stone to the world but did not give it a name. A chemist who conducted early analyses of the rock called it *dolomie* in Dolomieu's honor, and this soon became anglicized as "dolomite."[3]

By the early 1800s, geologists learned that dolomite was technically not a rock but a distinct *mineral*—a carbonate with equal amounts of calcium and magnesium, in a specific crystal form. But even today, geologists tend to use dolomite (the mineral) and dolostone (the rock made mainly of it) interchangeably—and the entire mountain range where Dolomieu first worked, the Dolomites, is now named for the rock named for the mineral that is named for a man who was named for his hometown.

Over the following decades, geologists mapped thick units of dolostone in stratigraphic sequences on every continent and found that it was in fact more abundant in the rock record than limestone. In the mid-twentieth century, dolostone became an important target for petroleum exploration, and geologists working in the oil and gas industries were the first to articulate the "dolomite problem."

Oil and gas can be extracted only from rocks that have pore spaces where it can accumulate, so petroleum geologists are obsessed with rock porosity and how it is formed. Dolostone almost always has microscopic pores in it, and in the oil business, understanding the origin of those tiny voids could mean better predictions about where to find productive fields—and make bigger profits. So, like good Lyellians, industry geologists began looking for modern marine settings where extensive banks of dolostone might be forming, and, to their frustration and bewilderment, found none.

Geologists had known for some time that the recrystallized texture of dolomite, often described as "sugary" or "sucrosic," indicated that it had started as one type of solid and then reorganized at the microscopic scale into another. And because dolostone was similar to limestone in every way other than its concentration of magnesium, they concluded that its formation must involve the passage of magnesium-bearing fluids through a precursory limestone, which then recrystallized to dolomite. So the hunt began for settings where newly formed limestones might interact with water high in magnesium.

Given that most limestones are marine, seawater was a logical candidate. Indeed, seawater typically has on average five times as much magnesium as calcium, which might lead one to think that dolomite could precipitate directly from the oceans. The fact that it is not spontaneously forming anywhere on the seafloor as a primary deposit suggests that much higher magnesium concentrations are needed for it to grow. Geologists developed several scenarios for how this might happen.

One of these is the "evaporative reflux model." When seawater evaporates, as it might in a nearshore lagoon, different minerals crystallize out in a specific sequence: first calcite, then gypsum, then rock salt. Because the precipitation of these minerals removes calcium and other ions—but not magnesium—from the water, the remaining brine would become progressively more enriched in magnesium. This brine would also be very dense and would tend to sink down into the subsurface, where it could infiltrate recently formed limestone, swap some of its magnesium ions for calcium—and form dolostone.

Lagoons like this do in fact occur in the Persian Gulf region, where wide, flat supratidal basins called "sabkhas" provide the ideal setting for concocting high-magnesium waters, and this afforded geologists studying the dolomite problem some Lyellian comfort. The evaporative reflux model also accounts for thick gypsum deposits that lie on top of some dolostones—as in the case of the Permian reef of Texas and New Mexico. In fact, the extensive cave system at Carlsbad Caverns National Park reflects this stratigraphic arrangement: sulfuric acid derived from water seeping through the gypsum has dissolved away the underlying carbonate rocks, forming the impressive caverns. But many dolostones, including the Cincinnatian Series in Ohio, are not associated with gypsum deposits—and so the evaporative reflux mechanism can't be the only way to make dolostone.

Some geologists think that the reason dolomite doesn't form directly in marine settings is simply the sluggish nature of dolomite crystal growth—what chemists call slow "reaction kinetics." Proponents of this view emphasize that dolomite formation is an incremental, multistep process that begins soon after deposition of the precursor limestone but may continue for millions of years as the sediments are buried to greater depths and encounter higher temperatures. Support for the "burial dolomite" model comes from the observation that the ratio of dolomite to limestone increases in the rock record as one goes back in time.[4] This model seemed most likely for the Cincinnatian Series, but I could never convey it persuasively to students because I felt I hadn't heard it directly from the rocks. I just couldn't understand what dolomite was saying.

When I started my position as assistant professor, there was no formal mentoring in how to teach. The expectation was that faculty members would figure it out on their own, and the implication was that teaching didn't really matter as much as scholarly productivity. As I became more comfortable in the classroom and more aware of what fostered—and impeded—student learning, I began to feel frustrated with the way geology was being taught, not just at my own university but also across the country. The field was changing in exciting ways, energized by new interdisciplinary approaches to thinking about Earth's complex tectonic, hydrological, and biological systems over geologic timescales. But the curriculum was still anchored in the logic of nineteenth-century museums, with separate courses on Earth history, mineralogy, petrology (study of rocks), geomorphology (study of landforms), structural geology (study of crustal deformation), and paleontology (with emphasis on fossil identification rather than on evolutionary patterns). The emerging fields of geochemistry, neotectonics, and, more important, climate science were simply absent, or at best, wedged in as footnotes on the last day of traditional courses. To me, it seemed obvious that the traditional course sequence no longer reflected the discipline, and although some members of my department agreed, there was little collective appetite for change.

My colleagues' indifference had an advantage: No one cared how I taught my own classes, and I was free to rethink their content and format. Introductory geology courses are typically a breathless race through a condensed version of the entire major. I began to realize that students actually learned more when I covered less. I changed my emphasis from teaching to the handful of students who might end up majoring in geology to focusing on all the other Earth citizens—fellow Earthlings—in the class. And I spent more time on the epistemology of geology—i.e., how we have come to know what we know about the planet.

We talked about the discovery of Deep Time in the late eighteenth century, the emergence of evolutionary thinking in the nineteenth, the plate tectonics revolution in the twentieth, and the growing awareness of biogeochemical cycles—the endless exchange of elements among rocks, water, and life—as one of Earth's key attributes. We discussed James Lovelock's provocative "Gaia hypothesis," that Earth might be considered a superorganism whose biogeochemical cycles are like the physiological processes of a living creature—and whether this was a viable scientific idea or simply a powerful metaphor. Covering the intellectual history of the geosciences humanized the field and opened a door for the science-phobic one-third of the class who enrolled just to fulfill a graduation requirement.

In my fourth year of teaching, I gave a presentation at a professional meeting about the revised lab manual I had developed for our introductory course. The talk caught the attention of an editor working on a nontraditional new geology textbook that was to be structured around global Earth systems rather than on geological taxonomies. The editor had been looking for a set of laboratory exercises that could be packaged with the textbook, and my homegrown manual seemed to be a good match. I was offered a contract to produce, within a year, a version of the manual that could be used not just in southern Ohio but also anywhere in the world.

The textbook's primary author was an eminent geologist at Yale, and the publisher arranged for us to meet during the Geological Society of America conference, which happened to be in Cincinnati later that year—close enough for me to drive there and back during daycare hours. I knew of the Yale professor's formidable reputation (and ego) but was emboldened by what I perceived to be our shared commitment to a more enlightened approach to teaching the geosciences. We met over lunch and reviewed the outline for the textbook. I noticed a chapter dedicated to global cycles of carbon, phosphorous, and nitrogen. The content paralleled a section in my lab manual, and I commented that the manual complemented the textbook nicely on this topic. The Yale professor seemed pleased about that. Feeling more confident, I mentioned that at the close of the element cycling section, my manual provided a short exposition of the "Gaia" idea and posed some questions meant to engender thoughtful discussion about it. The professor suddenly looked dyspeptic. He replied sharply, his voice rising above the ambient conversational murmur in the restaurant, "Such pseudoscience has no place in my work!" I wasn't completely surprised—"Gaia" had New Age-y overtones—but the vehemence of his reaction still strikes me as interesting; what was so threatening, so subversive, about having students just consider the idea that Earth could in some sense be alive?

Three decades on, geoscientists still hesitate to use the term "Gaia" to describe the many complexly intertwined cycles that are the signature habit of Earth. But the basic premise of Lovelock's once-controversial idea is now mainstream: There is no bright line between the living and nonliving components of the Earth system. Everywhere on Earth, at every spatial and temporal scale, rock, water, air, and life are in intimate communion. "Gaia" has evolved into the burgeoning field of biogeochemistry, with its own graduate programs, professional societies, and specialized journals.

A biogeochemical lens also provided a breakthrough in the long-running case of the "dolomite problem." As early as the 1950s, some geologists had noted a correspondence between the degree of dolomitization and the presence of organic matter in limestone, but they had no idea what the causal mechanism might be. In the late 1980s, researchers supported by oil company funding began to publish papers suggesting that bacterial reduction of sulfur compounds in sediments seemed to be linked with modern dolomite formation.[5] Still, many in the dolomite community remained skeptical of this "microbial model" because the exact geochemical mechanism remained elusive. Also, at sites where bacteria seemed to be fostering dolomite precipitation, further investigation showed that the mineral was actually an impostor—a less-ordered form of calcium-magnesium carbonate, not true dolomite.[6]

More recent, however, genuine dolomite has at last been grown in the lab at low temperature, on biofilms—less technically, slime—created by certain sulfur-processing bacteria in the presence of manganese, a common trace element in seawater and groundwater.[7] The ooze secreted by these organisms, it seems, catalyzes the otherwise reluctant dolomite-forming reaction. In other words, teeming microbes buried in limestone sediments would appear to be the secret crew behind Dolomieu's acid-resistant rock. In partnership with the tiny calcifying organisms that formed the progenitor limestone, they are the architects of the great dolostone platforms that have kept most of the planet's carbon safely locked up for hundreds of millions of years. The dolomite problem has yet not been completely resolved, but it seems clear that microbes are centrally involved.

This consensus is part of a larger paradigm shift in mineralogy, which was previously the "tidiest" subdiscipline in the geosciences, a kind of sterile Platonic realm governed purely by the laws of chemical bonding and crystal geometry. Just as the geosciences as a whole had adopted a global systems perspective with a new understanding of plate tectonics and biogeochemical cycling, mineralogy too began to embrace a more dynamic, process-oriented approach. Today, mineralogists recognize that more than 40 percent of all mineral species on Earth are in some sense biogenic—produced either directly or indirectly through the action of lifeforms— and that as life has evolved on Earth, so have the minerals in the crust.[8] Animal, vegetable, mineral, and microbe are very untidily entangled; Gaia is a stranger and more complicated creature than we ever imagined.

With no time during the academic year to focus on the manuscript for the published version of the lab manual, I had to work feverishly on it during the following summer. Suspecting that the Yale professor wouldn't actually look too closely at its contents, I did mention the Gaia hypothesis in the section on biogeochemistry. (It seems he never noticed.) I managed to finish the manual just before my second son, sweet baby F, was born in the sweltering days of late August. I had just earned tenure, and this time, I took my rightful maternity leave, savoring unscheduled days with my *bambini*. For the first time in years, I paused long enough to contemplate what the future might hold.

After six years, I still hadn't been able to set down roots in the Ohio dolostone. The land seemed trampled and defeated, and I felt that I was becoming similarly flattened and dulled. I wanted the boys to grow up feeling at home in nature and often took them to a park near our house where a creek cut through a bank of glacial till (remarkably, Pleistocene glaciers had even reached southern Ohio). Amid the broken glass and crumpled beer cans in the creek bed, I would find small pieces of gneiss and schist from the Canadian Shield—the ancient core of the continent—that I clung to like long-lost friends. I missed the feeling that wild, unpopulated places weren't too far away. And I longed to be closer to rocks that spoke more directly to my heart.

The lack of deformed rocks for hundreds of miles in any direction also posed considerable practical challenges; teaching courses about mountain building in a place without a hint of tectonic unrest meant that field trips for my classes were multiday, logistically elaborate affairs involving long drives and all the complications of camping with a large group. When my babies were still nursing, each needed to come along with me, strapped into a car seat in the university van and onto my back when we examined outcrops. Exhausting as these trips were, however, they were also full of joy; students doted on the kids and were endearingly protective of me. And they led to encounters that would have made university insurance officers swoon.

Once, in the wilds of West Virginia, pre-GPS, we took a wrong turn and got lost on back roads in a deep, heavily wooded valley. Rather than getting further confused, I decided to stop and ask for directions. I spotted a house, pulled into the driveway, hopped out of the van, and was walking toward the doorstep when a man came out with a shotgun. I heard students in the van scream, which startled me more than the confrontation itself; the tableau had seemed too much of a stereotype to be real. When the man saw I was not whatever antagonist he had been expecting, he lowered the gun, asked what the hell we wanted, and told us gruffly how to get back to the main road. As I climbed back into the van, the students, in a touching role reversal, rushed to comfort me.

Another time, we were running out of daylight and wanted to get to our campsite in the panhandle of Maryland on the opposite side of the upper Potomac River before dark. Even though the campground was only about five miles away, we needed to go twenty miles downriver to the next bridge, then another twenty back upstream again, by which time the sun would have set. One of my graduate assistants pointed out a small annotation on the road map close to where we were: "Toll Bridge." I was surprised, since this was a remote area far from any major highways, but agreed it was worth investigating. We arrived at the spot on the south bank of the river, and found that the "bridge" was a floating platform of plywood laid over a raft of inner tubes lashed together with ropes. The young man collecting tolls belonged to the family that owned the bridge—and rebuilt it, he added, reassuringly, each year after it was washed out by spring floods. We watched with skepticism as another member of the clan demonstrated the structure's trustworthiness by driving his pickup across and back. There were thirty undergraduates in three vans, and I had them get out and walk over this engineering marvel, which bobbed alarmingly under their footfalls. Then one by one, my graduate students, and I, with baby F in his car seat, made the trepidatious crossing with the vans in the deepening dusk. That evening in camp, under a starry sky, everyone was in a festive mood, having shared a memorable adventure.

These trips away felt liberating, a reminder of my predolostone audacity, my secret Arctic explorer identity. Geographically, the central Appalachians were the most logical place for me to establish a long-term research project with graduate students, but geologically, they were already claimed: There were plenty of academics at East Coast universities who had done their graduate work in the region, had far deeper knowledge of it than I did, and—as the likely reviewers for any grant proposals seeking funds for work in the area—would make it difficult for a newcomer to move into their territory. Projects based farther away would mean longer periods of travel, now with two children in tow. The prospect of continuing to scramble for graduate students and research money in structural geology, while based in a place where strata of undeformed dolostone stretched beyond the horizon, began to seem exhausting.

In our personal lives, it felt like N and I were out of equilibrium with our surroundings. We were socially isolated; most people in his circle were well past the point of having small children underfoot, and the few friends I had made were uncomfortable with our age gap. In any case, I was almost always too busy with work and children to be a dependable friend.

One nascent friendship with a young woman on the faculty in the humanities became chilly after a conversation in which she asserted—following the "deconstructionist" literary theory of Jacques Derrida, which was popular at the time—that all of science was simply a social construct, hopelessly tainted by ideological bias and political agendas, and that none of it actually described anything real or true. Far less practiced in philosophical discourse than she, I found myself spluttering about the irrefutable truths embodied in rocks, while she replied with rhetorical flourishes that we really have no way of knowing what they mean or signify. Also, she suggested, my involvement with the scientific establishment made me complicit in perpetuating a centuries-old patriarchy. For her, the conversation was probably just an intellectual game, but to me it seemed personal, a snide dismissal of things I cared about deeply. Feeling stung, I let our connection lapse over my loyalty to rocks. Deconstructionism, followed to its extreme, is a nihilistic scorched-earth philosophy that leaves nothing standing in its wake. Within the following decade, it had played itself out in academic circles, but ironically—given its roots in Marxist and feminist thinking—aspects of it can now be seen in the science-denialism of the far right. And for me personally, it damaged a promising relationship.

When we were in Ohio, reactionary forces were already stirring not far from us in northern Kentucky, where a fundamentalist Christian group called Answers in Genesis announced plans to open a tourist attraction called the Creation Museum, dedicated to denouncing geologic understanding of the age and evolution of Earth. Like the deconstructionists, the fundamentalists blithely ignored the authority of natural history. The fact that the museum would be built directly on top of the Cincinnatian Series, one of the best fossil records of Ordovician biodiversity, was an irony not appreciated by the creationist entrepreneurs. Our own charming college town, where idealistic civil rights activists were once trained, also had a retrograde side. The year I moved there, the KKK had rallied in defense of two high school boys who were expelled after dressing as Klansmen for a Halloween event. And when I had finally garnered the courage to submit a letter to the campus newspaper suggesting that the university drop its racist mascot, I received threatening phone calls and found a profane note on my office door.

Although there were many enlightened people in that community, it seemed to me that they were always on the defensive, under siege. I didn't want my children to grow up in such an environment. And I yearned for them to know their grandparents, pine forests, snowy winters, Lake Superior. N, too, was ready to leave southern Ohio, and we started to think about an exit strategy. He was old enough to take advantage of a generous early retirement plan, and I began to look for job openings farther north. The options were limited. Unless departments are seeking an experienced faculty member to serve as a department head, virtually all academic job postings are for new assistant professors. And a new start somewhere else carried a risk: I would have to earn tenure, again.

I decided to apply for an assistant professor position at a small liberal arts college, Lawrence University, in northeastern Wisconsin, suspecting that my file would be dismissed immediately because I was already a tenured associate professor. Instead, I was invited for an interview. The first thing I noticed about the campus, with mild dismay, was that many of its buildings were clad in an all-too-familiar rock: dolostone, from the nearby Niagara Escarpment. But glacial boulders of gneiss and granite scattered around the grounds were reassuring signs that the Canadian Shield was close. The dean offered me the job, and although it meant a cut in salary, I accepted. I would be appointed as an associate professor, the rank I already held, but without tenure. My tenure evaluation would occur after two years. If I had known what the next two years would bring, I might have been concerned. But I was sure that better times lay ahead.

N and I thought that we had now set the stage for a simpler, happier life. He would look after the boys part-time, but there was an excellent daycare center near campus, so he would also be free to continue to do research. My class sizes would be considerably smaller, and I imagined I would have less grading to do. Because Lawrence was an undergraduate institution, I wouldn't have to recruit and advise graduate students and seek external grants to support them. Since I would be the primary working spouse, we hoped to be viewed by our new community on more equal footing. My parents would be within a few hours' drive. Everything would be easier.

This was a lovely fantasy. We had escaped the doldrums, the dolor, the stubborn problem of dolomite, but we could not escape the immutable fact of Time.

GRANITE

On a muggy afternoon in June, I am in the basement of a once-grand campus building, despairing at the disarray around me: stacks of moving boxes, a wall of half-emptied specimen cabinets, a heap of scrap metal, piles of musty stuff bound for the dumpster. I feel a little dizzy, overwhelmed by the heavy air and the enormity of the task we face. Spring term has just ended, and I've still got final exams to grade, but two student workers and I have three weeks to pack up everything in the geology department before this creaky old building is demolished to make way for a modern new science complex. The geology department has been in this space since the 1890s, and we are faced with one hundred years of accumulated objects: lab instruments, glassware, rock saws, sediment sieves, file cabinets, field equipment, camping gear, microscopes, projectors, maps, books—and *so* many rock samples.

It's grueling work, both physically and emotionally—there are ghosts here among the century of artifacts, glimpses of entire faculty careers in earnest course notes and visual aids of earlier times. Do these things merit preservation? I bless them with a silent word of acknowledgment before adding them to the trash heap. We are in triage mode, making draconian decisions about what to save and what to jettison, when one of the students moves a display cabinet and discovers a door behind it. It leads to a subbasement we didn't know existed, with a ceiling so low we have to crouch to enter. The space is filled with rocks in wood crates, collected by students and faculty on long-ago field trips mainly in northern Wisconsin and the Upper Peninsula of Michigan. We haul a few of the boxes up to the main level and pry them open. The crates seem ancient, but the rocks are fresh as ever: "red granite" (Wisconsin's official state rock), porphyritic granite, two-mica granite, granite with inclusions, granite pegmatite, granitic gneiss—so many variations on a theme of pink, white, and black, each a chapter in the story of how the continent was built. I can guess at the specific outcrops that many of them came from and feel a little sad on their behalf, being locked up in a mildewy basement since who knows when. My absurd empathy for a box of rocks distracts me for a moment from the colossal amount of work ahead of us.

The three years since I took the position on this campus had been tumultuous. Among other upheavals, my department had evaporated around me. When I arrived, I had had two colleagues, a young sedimentologist (a dolostone specialist, in fact) and an older, jack-of-all-trades geologist who taught mainly introductory courses and acted as the department's technician—while also serving as the university's assistant football coach. In my first year, the younger colleague was denied tenure by a university-level committee, and it was too late in the academic year to launch a search for his replacement. A few months later, my older colleague announced that he was ready to retire. Suddenly, I was the only surviving member of the department and in a vulnerable situation. I wasn't yet tenured in my new position, and I knew that members of the other science departments did not hold the geology program in high esteem. There was a real possibility that the administration would view the rare circumstance of two vacant positions as an opportunity to close down our department and reallocate the tenure lines to other programs.

Meanwhile, any hopes for a simpler, more balanced home life were in shambles. Although I was now an experienced teacher, the move from a research university with more than a dozen departmental colleagues to a tiny liberal arts college with a department of three—and then just one—required me to develop courses across a much wider range of subjects, some well outside my own area of training. These included courses on igneous and metamorphic rocks, Earth history and paleontology, as well as climate science and environmental geology. The university also had a "great books" first-year seminar that all faculty were to teach with reasonable proficiency as a condition for tenure. That first year, the reading list included John Locke's *Treatise on Government*, Friedrich Engels's *Socialism: Utopian and Scientific*, Sigmund Freud's *Civilization and Its Discontents*, and Hannah Arendt's *Eichmann in Jerusalem: A Report on the Banality of Evil*, among other frothy titles. We were expected to guide eighteen-year-olds expertly through these tomes and elicit from them thoughtful discussion and insightful essays—a daunting challenge, since I was reading them for the first time myself.

I could not have foreseen the collapse of the department, but I had underestimated how much time so much new course preparation would take. I was exhausted all the time and irritated at the disparity between the overwhelming demands I faced while my husband enjoyed unscheduled days. He had found being home with the children too difficult, and we had enrolled them in full-time daycare.

The reality of our age difference was becoming too great to ignore. We had naively believed that our strong intellectual affinity canceled out the inescapable fact that we belonged to different generations. When we first talked about getting married, I had thought N and I had parity in making the decision. But all these years later, having reached the age he was then, it's obvious to me that we did not come into the marriage with equal understanding of the three decades that divided us. He could remember being in his late twenties, but I could not then imagine being in my midfifties. We geologists may think we have special insight into the passage of time, but, in fact, one can understand the real meaning of time only by living it.

N and I were at the point of an amicable separation and starting to envision how to reconfigure our lives when I discovered I was pregnant again, with a third son. And then, a few months later, N was diagnosed with an aggressive form of leukemia. His chronic fatigue—and consequent impatience with the boys—had probably been premonitory signs. And so, in that summer when the old science building was being demolished, I found myself caring for three small children, including newborn K, and a terminally ill husband.

Against this backdrop, the process of moving the entire inventory of the geology department was almost therapeutic. It provided a mental respite from the nearly unbearable stresses of our family life. Sifting through all the relics left by my predecessors also constituted a deep dive into the intellectual history of geology.

Geology was one of the original subjects (together with Latin, Greek, German, mathematics, and philosophy) Lawrence University offered at its founding in 1847. The science was still in its formative stages at that time. Lyell's *Principles of Geology* had been published in just the previous decade, and the notion of a global Ice Age—the eventual key to understanding Wisconsin's landscapes—was a controversial new hypothesis. Darwin's *On the Origin of Species* would not appear for another dozen years. The basic divisions of the modern geologic timescale were still being worked out, gradually replacing an old four-part system (Primary, Secondary, Tertiary, Quaternary) that linked certain rock types with distinct intervals in time. In this scheme, granite was a Primary rock, formed only in primeval times, and the "basement" upon which Secondary, Tertiary, and Quaternary rocks were deposited. As we lugged the crates of granite up out of the cellar, it occurred to me that their subterranean storage was stratigraphically appropriate, according to that logic.

The origin of granite has been one of the longest-running debates in geology. As the foundation for the continents, it's both important and abundant—but it's also deeply mysterious. Granite is an invention apparently unique to Earth in our solar system, even though all of the rocky planets probably started out with the same raw materials and have the necessary ingredients. In other words, Earth is the granite planet, and the question of the genesis of granite is thus fundamental to elucidating how the planet works. The history of geologic thinking about granite illustrates the push and pull between the peephole views that individual rock outcrops provide and the panoramic theories that govern the way geologists interpret them. The path toward understanding granite has been long and meandering, looping back on itself over the course of two hundred years.

In the late eighteenth century, a prevailing school of

thought called "Neptunism" held that all rocks—including granites—were deposits from seawater, formed either by physical sedimentation or precipitation from solution. (The name was an allusion to the god of the seas, not the planet, which wasn't discovered until 1846.) The leader of the Neptunists, a kind of Captain Nemo who viewed the world from the submarine realm, was the charismatic German mining geologist Abraham Gottlob Werner, born around 1750. He had an encyclopedic knowledge of mineralogy and made important observations about the geologic "habitats" of mineral ores, but his obdurate belief in the marine origin of all rocks blinded him to other fundamental geologic facts and led to elaborate subhypotheses that his devout followers pursued unproductively.

Interpreting granite as a marine deposit, for example, required that the primordial seas had a composition utterly different from that of modern oceans. The Neptunists, who didn't believe in igneous rocks, also needed to account for the very real phenomenon of volcanism, which Italian geologists—quite reasonably—presented as prima facie evidence against the doctrine. The 1805 eruption of Mount Vesuvius was enough to change the mind of one notable Werner acolyte, Alexander von Humboldt, who fell from the Neptunist faith after witnessing that event.[1] Werner and his still-faithful Neptunist disciples were not impressed, however. They admitted that under some special circumstances, rock could be melted, but they considered magmatic activity to be a shallow phenomenon caused by coal seams burning underground. The Neptunists did not even recognize basalt as an igneous rock, much less the great massifs of granite that formed the shield regions of the continents.

If Werner was a zealot obsessed with water as the source of all rocks, his contemporary James Hutton—the versatile Enlightenment-era Scotsman credited with discovering Deep Time—was his counterpoint, obsessed with fire. His interpretation of the erosional unconformity at Siccar Point, Scotland—with its tilted turbidites overlain by undeformed sandstones—was based on his theory of a self-renewing Earth fueled by a great internal heat engine. Long before any of his contemporaries, Hutton intuitively grasped the concept of the rock cycle, in which preexisting rocks may be eroded to form sedimentary strata, or contorted in mountain building to form metamorphic rocks, or melted to form igneous bodies, in an endless process of reincarnation—a prescient glimmer of plate tectonics—driven by a mighty forge in the heart of the planet.

Hutton was right about so many things, but he was not exactly a paragon of objective, dispassionate science. He had formulated his grand theory first and was constantly looking for corroboration of his scheme—which, as any middle schooler could tell you, is a flagrant violation of the scientific method. Today, most people encounter Hutton's ideas as they were transcribed by his friend John Playfair, another intellectually omnivorous Scotsman, who was a generation younger. In a memoir written in 1822, a quarter century after Hutton's death, Playfair recounted an outing in Scotland's Cairngorm Mountains, where a positively giddy Hutton found confirmation of his heat engine hypothesis.

In outcrops along the River Tilt, they came across fingers of pink granite that had made inroads into dark metasedimentary rocks in a manner that could not be explained unless the granitic material had been molten and fluid at the time—and had come up from below rather than being deposited from above. This cross-cutting relationship also showed that granites were not invariably primordial but, in some cases, younger than other rocks. Playfair describes the eureka moment:

> In the bed of the river, many veins of red granite ... were seen traversing the black micaceous schistus. . . . The sight of objects which verified at once so many important conclusions in [Hutton's] system, filled him with delight; and as his feelings, on such occasions, were always strongly expressed, the guides who accompanied him were convinced that it must be nothing less than the discovery of a vein of silver or gold, that could call forth such strong marks of joy and exultation.[2]

We get a glimpse here of Hutton as a single-minded fanatic; but the fact remains that he got very close to the truth in a way that many geological ideologues before and after him did not.

Also, I can attest from personal experience that most geologists have whooped in exactly the same way upon discovering rocks that confirmed a partly formed idea. For me, this anecdote, more than any other, brings Hutton fully to life. I recognize his pure joy in having grasped the logic of Earth. I've experienced it with diamictites in Svalbard and turbidites in Ellesmere, and shared it with students at outcrops, roadcuts, and quarries—the thrill that comes from being able to read the raw rock record.

In my experience, the best geologists have a subliminal connection with rocks. When one spends enough time in their company, rocks have the power to seep into one's subconscious. I suspect that Hutton's grand Theory of the Earth did not spring fully formed in a flash of insight but had been taking shape in his mind, perhaps without his even being aware of it, as he ambled about his land on the Scottish borders, observing soils, streams, and seaside cliffs, year after year. Over time, he gained a sense for the ways of the landscape— and developed, in biologist Barbara McClintock's words, "a feeling for the organism"[3] that wasn't based on any one observation but was instead the aggregate effect of many geological outings on his receptive mind.

So while it is true that Hutton had already formulated the notion of Deep Time before he saw Siccar Point, and the idea of igneous intrusion before visiting Glen Tilt, those concepts surely came from internalizing many earlier observations. If he did not follow the middle-school version of "the" scientific method, it is because that method is poorly suited to the study of Earth. Hutton somehow had the capacity, decades before the practice of geology became systematized, to conceive of phenomena at multiple scales, to move inductively from an outcrop to the planet as a whole.

From about 1820 to 1870, geology matured from a pastime practiced by self-taught amateurs like Hutton and Playfair into a formal discipline with academic hierarchies and professional societies serving as gatekeepers. During this time, geologists focused on the enormous task of inventorying the planet— naming, describing, and cataloging rocks, minerals, fossils, and landforms around the globe. More telling, this period produced few geologists as creative and insightful as Hutton, whose mind had not been shackled by pedantic classification schemes.

In the 1850s a new analytical tool, the petrographic microscope, revolutionized the study of rocks. A British geologist named Henry Sorby found that when rocks are cut into very thin slices (about 1/1000 of an inch), they will transmit light, but the light that passes through them is refracted and polarized according to the crystal structures of the minerals within them. When various filters are placed between a back-illuminated thin section and the lenses of a microscope, different mineral species will appear to stand at different "heights"—and will display varied, vivid colors that change kaleidoscopically as one turns the microscope stage. This is mesmerizing and trippy—like a psychedelic, cinematic stained glass window. The sheer beauty of rocks in thin section was in fact one of the things that convinced me to become a geologist.

More important, the color and other properties of the light that passes through crystals make it possible to determine a rock's mineralogical composition. Thin section examination

A microscopic view of granite shows interlocking crystals of quartz and feldspar. Field of view is one quarter inch in diameter.

also reveals the "texture" of rocks—the nature of boundaries between crystals of different minerals, which bears information about how they formed. Similarities with the texture of industrial metals, quenched from melts, convinced most geologists that granites had indeed crystallized from the molten state. By 1900, Hutton's intuitive concept of granite as a solidified magma was accepted as fact by most geologists, even though his larger vision of a vigorous, self-rejuvenating Earth slowly faded from view.

In the early twentieth century, however, the question of granite's origin was reopened by a new generation who saw that the magmatic theory led to unresolved conundrums. First, as geochemical methods improved, a growing array of igneous rock types had been identified, raising the question of how there could be so many distinct magma sources in the Earth's interior. Some geologists thought that the mantle was a "plum pudding" with pockets of different composition, including distinct spots from which granitic magmas emanated, but there was no direct evidence or explanation for such heterogeneity. At the same time, field geologists pointed out a geometric puzzle that became known as the "room problem": If massive granite bodies, or "batholiths" like those in the Sierra Nevada, represented magma intruded into other rocks, how had space been made for so much magma—and what had become of the rocks it intruded?

To some geologists, particularly in Scandinavia and Britain, the obvious answer to both riddles was that granites were not in fact magmatic intrusions but instead formed in place by the chemical conversion of sedimentary rocks in a cryptic process they called "granitization." Adherents of this idea disagreed about the details of this process, but it was thought to be broadly akin to the way in which petrified wood forms, as the original cellulose is replaced atom by atom by ions dissolved in groundwater. Today, granitization is seen as an embarrassing detour in the intellectual development of geology, and Hutton would surely have been dismayed to learn it gained traction more than a century after his exultant discovery in Glen Tilt. But understanding the granitization delirium is important because it illustrates how scientific, political, and personal factors can converge in ways that lead to spurious consensus. In the case of granitization, nationalism, egotism, and misogyny all played a part.

The granitization hypothesis germinated from the accurate observation that "host" rocks in the vicinity of granites are commonly altered to the extent that the precise boundary between them and the granite cannot be discerned. Around 1900, French geologists working in the Alps and the Pyrenees surmised, correctly, that heated fluids emanating from the granitic bodies had probably moved into the surrounding rock mass and caused these profound changes. In the 1930s, Swedish geologists working in the Baltic Shield—the old geologic center of Scandinavia, where granites and granitic gneisses abound—began to take this hypothesis further. They pointed to sites where granite bodies not only graded imperceptibly into sedimentary rocks but also had sheetlike forms that resembled bedding planes, the surfaces between layers of stratified rock. Moreover, these granites did not appear to be connected to any fractures or conduits that could have conveyed magma up from depth. To the Swedes, the logical interpretation was that the granite had not been introduced as a magma from a deeper source but had been formed in place via chemical alteration of the original sedimentary rocks. This was the beginning of a mass delusion that shaped and skewed entire careers.

Inspired by the radical reinterpretation of granites in Sweden, British geologists working in the Scottish Highlands began to identify similar sites where it appeared that granites had formed through modification of preexisting rocks. It is possible that the hypothesis was especially attractive in the United Kingdom in those interwar years because German geologists were staunch magmatists.[4] In any case, some of the most revered British geologists embraced granitization. These included Arthur Holmes, a brilliant scientist who, in 1911 at age twenty-one, obtained the first quantitative ages for rocks, just a few years after the discovery of radioactive decay. Holmes's later work showed that Earth had to be at least 3 billion years old, at a time when astrophysicists, based on imprecise estimates of the Hubble constant, asserted that the *Universe* was only 1.8 billion years old. Holmes also posited an early version of plate tectonics theory decades before other geologists were ready to embrace the idea of drifting continents.[5]

After his first wife died, Holmes married one of the few prominent female geologists in Britain, Doris Reynolds. She was a vigorous, energetic person and a well-established scientist who had taught at University College London and Durham University before moving with Holmes to the University of Edinburgh. Unfortunately, her legacy is marred by her obstinate adherence to granitization.

Reynolds and other "transformists" saw themselves as scientific revolutionaries, shaking up old paradigms and forcing geologists to think about rocks in fresh ways. Indeed, for a time in the late 1940s, the argument for a cold origin of granite was the hottest topic in geology. By the late 1950s, however, it was a fading theory undone by internal inconsistencies. Like a religion splintering into sects, the transformists had broken into factions with different views on the actual mechanism of granitization. Some thought it was a dry process of element migration, driven by heat and chemical gradients. Others, the self-named "soaks," insisted it was wet, caused by mineralizing fluids seeping through rocks.[6] It didn't help that some in the latter camp referred to these fluids as "ichor," after the substance that was said to flow in the veins of Greek gods. Furthermore, new experimental work on magmas was resolving some of the questions that had motivated the hypothesis in the first place. But Reynolds, a dry diffusionist, was undeterred. When critics argued that granitizers could not identify the source of the "emanations" that supposedly transformed sediments to granite, she pointed out that the great Hutton himself had invoked heat welling up from within the Earth without having any idea of where that heat came from.

I have to imagine that for Reynolds, any reasonable objections to her ideas were difficult to disentangle from misogynist attacks—and what she memorably called "male authoritarian bluff."[7] I suspect that some of her stubborn adherence to granitization was a survival skill learned through years of having had her views summarily dismissed by the academic patriarchy. I wish for her, retroactively, that she had had the luxury of admitting she'd simply been wrong and had then been free to move on and investigate other things. But when cramped long enough by fear of ridicule, people become permanently bent, no longer able to unfold into their full potential.

The commotion over granitization had ended before I was born, but through my husband, N, I had a personal window into that epoch in the history of geology. When N was a new graduate student at the University of Edinburgh in the late 1950s, Holmes had just retired, but Reynolds was looking for PhD students to shore up the crumbling citadel of granitization. To Reynolds's credit, she did make specific predictions based on her theories that could be tested and potentially falsified—the essential criterion for scientific inquiry. N became one of her last students, assigned to a project on the Donegal granite in Northern Ireland. He was to look for chemical "fronts" of iron and magnesium in the surrounding rocks, which, according to Reynolds's theory of granitization, would have been expelled from the rocks as they were transformed to granite. After several field seasons of mapping and sampling, he found no such fronts—and finally concluded that the granite had formed from a cooling magma. N was a deferential, well-mannered person, and I can only imagine the awkward scene at his dissertation defense when he presented evidence that contradicted the life work of his advisor.

In the 1920s, even before the peak of the granitization frenzy, an experimental geochemist at the Carnegie Institution in Washington, DC, Norman Bowen, had in fact solved one of the major questions underlying the "granite problem": how Earth has produced such a wide range of igneous rock types. At the time, the "plum-pudding" model of the Earth's mantle, which imagined distinct reservoirs of different composition, was being disproven by new insights about Earth's deep interior inferred from the propagation of seismic waves through the planet following great earthquakes. It was becoming clear that the mantle was relatively uniform in composition, consisting of the rock type called peridotite, which is made primarily of the iron- and magnesium-rich silicate mineral olivine. This meant that somehow, all the other rock types on Earth—including granite, which has no olivine and comparatively little iron or magnesium—had been distilled from the same global "parental" mantle material.

Beginning in the 1920s, Bowen created batches of "mantle" magma in the lab by melting peridotites and observing them as they cooled and crystallized again. Critically, he showed that different minerals crystallize at different temperatures, and that if early formed, high-temperature minerals like olivine are removed from the remaining magma—which in nature could happen by crystals settling out under the influence of gravity—then the residual melt will evolve toward compositions far different from the bulk chemistry of the original. This is conceptually similar to the process of distillation of crude oil at refineries in which various types of hydrocarbons like diesel oil, kerosene, and gasoline are separated from the "parent" crude through their different boiling points.

Bowen found that if he stopped this "fractional crystallization" process at an early stage, after only olivine had crystallized, a melt of basaltic composition would be left. But if he continued to cool the magma and systematically remove crystals of later-forming minerals, a small volume of granitic melt remained. The essence of Bowen's work was that although the mantle is far from granitic in composition, it contains all the ingredients for granite, and fractional crystallization was the key to extracting them.

Fractional crystallization is now one of the central concepts taught in introductory geology courses. The sequence in which minerals crystallize from a mantle melt with decreasing temperature is often called "Bowen's reaction series," and it is a geologic shibboleth that once unexpectedly helped me through US Customs. I was coming back to the United States from Svalbard one summer, with my luggage full of precious rock specimens, and landed at Chicago's O'Hare airport to an overcrowded arrivals hall. Nervous about having my rocks impounded by overzealous inspectors on the lookout for agricultural pathogens, I tried to carry my bags in a way that didn't hint at their weighty contents. With luck, I thought, given the huge number of people to be processed, I might be waved past the inspection station. But the customs agent beckoned to me, hefted my suitcases onto his table, and asked what on Earth was in them—rocks? "Yes, rocks," I replied, with a sinking feeling, "I'm a geologist." As I braced myself for the possibility that all my work could be put on hold for months while the specimens were in bureaucratic limbo, he said, "OK, then tell me Bowen's reaction series." Like an automaton, I blurted out: "Olivine, augite, hornblende, biotite, orthoclase, muscovite, quartz!" He handed my bags back, and my rocks and I were free to go.

Introductory geology students—including, I presume, my customs agent—may memorize the sequence of mineral crystallization from a cooling magma without fully grasping its profound implications. If, as Bowen's work showed, the full spectrum of igneous rock types—including granite—can be generated by the process of fractional crystallization, there is no need to invoke a mottled mantle pudding of heterogeneous composition; all igneous rocks—and by extension, all sedimentary and metamorphic rocks as well—can trace their lineages back to the same ancestral mantle material. This is a foundational assumption of modern geochemistry.

Furthermore, Bowen's experiments demonstrated that the bulk composition of most granites, with about equal parts quartz and sodium or potassium feldspar (orthoclase), is close to a "eutectic" or "minimum melt" composition—that is, the lowest temperature melt that can exist for a given combination of ingredients. This was strong confirmation that granites came from magmas, but it also had implications that Bowen himself seemed not to fully grasp and that would not be appreciated until plate tectonics provided an explanation for why rocks melt at all.

The granitizers were well aware of Bowen's work, but as field geologists, most of them were skeptical about the relevance of idealized lab-based research in explaining complicated phenomena in the natural world. And Bowen's results did not address the "room problem," which loomed large in the minds of the granitizers. (The consensus today is that the "host" rocks are domed up by, and, to a lesser degree, assimilated into, intruding granitic melts.)

Bowen sparred openly with another prominent granitizer, H. H. Read, who as president of the Geological Society of London wielded considerable power in international geological circles. Today, Read is largely forgotten, but his mantra that "the best geologist is the one who has seen the most rocks" survives, often emblazoned on geology club T-shirts (few who cite the quote realize the irony in it, given Read's strenuous advocacy of a discredited theory). In 1948, Read used the occasion of a guest address to the Geological Society of America to make sarcastic comments about Bowen and his work. Tragically, Bowen died by suicide in 1956, just as the emperors of granitization were being shown to have no clothes. The circumstances are not clear, but the public denigration he suffered must have exacted a psychological toll.[8] The stakes in the granite controversy were high.

By the late 1960s, granitization had been abandoned, and the new paradigm of plate tectonics had at last provided a framework for where and why magmas form. Since then, geologists have also attained a subtler view of how Bowen's results actually apply in nature. We now realize that granite is unlikely to be extracted in a single step directly from the mantle in the way that Bowen emphasized. Instead, his observation that granites represent "minimum melts" is considered a key clue that most granitic magmas probably come not from fractional crystallization of primary mantle magmas, but from fractional (i.e., partial) *melting* of preexisting rocks that have already been refined through multiple cycles of fractional crystallization. Bowen's observation that different minerals crystallize at different temperatures is the key, but the full spectrum of igneous rock types is more likely to have been created over time by "sweating" melts from solid rock rather than from minerals being plucked out of magmas. From his vantage point in the laboratory, Bowen hadn't appreciated the role of geologic time—and Hutton's indefatigable rock cycle—in the process of distilling granite from the mantle.

As we open still more granite-filled crates in the subbasement recesses of our doomed building, I think about the paradigms that prevailed when they were collected. The rocks have remained as they always were, while our interpretations of them have vacillated and evolved. The same is true for the events of our lives; the past is immutable, but its meaning changes with time.

I remember that terrible year as a maelstrom of exhaustion, anxiety, guilt, and despair. We managed to move everything worth saving from the old building, although tons of rock specimens remained in the basement when the wrecking ball arrived. The next task was to reconstitute the geology department physically—and repopulate it with faculty colleagues. We would be in temporary quarters for three years, while a new campus building was being constructed on the site of our old one and an adjacent building was being renovated, during which I would interview and hire two new colleagues.

Meanwhile, I was staggering under the weight of my dual roles as mother to three small boys and nurse to an estranged, terminally ill husband. His cancer was a malign and demanding lodger who had arrived uninvited then dictated everything we did. The knot of emotions I felt made it difficult for me to care for N without resentment and to help the boys process their own feelings about their father's illness and eventual death. My inner reserves of energy were utterly depleted; there was nothing more to draw on.

In the numb months after N died, it was almost a shock to realize that I was still in my midthirties. I was worn down, desperately sleep-deprived, haunted by the lingering specter of terminal illness. But daily life with the children tumbled on. There was no possibility of pausing to grapple with the enormity of what had happened to us. There were meals to be made, laundry to do, baths to be given, bedtime stories to be read, and once the boys were sleeping, papers and exams to grade.

We were still relative newcomers to the community, and I had not had the luxury of any free time to establish deep social connections. People we knew expressed condolences but were unsure of whether it was appropriate to offer help. In any case, I would probably have declined, out of Scandinavian stubbornness and a sense that I didn't deserve sympathy. I didn't realize how fragile and alone I really felt until one day when I picked the kids up from daycare. Three of the teachers, all Hmong women in their fifties who had come to the United States as refugees after the Vietnam War, presented me with a huge bag of used clothes for the boys, gathered for us from many households. To these women, the loss of a father and husband was the greatest calamity that could befall a family. They understood that we needed help, whether we asked for it or not. I crumpled onto one of the toddler-size chairs and cried.

My parents retired early and prepared to sell the house in the woods that my father had so lovingly built, in order to move across the state to be near us. When a house down the block from us came up for sale, they bought it. The boys would be able to run freely between mother and grandparents and grow up with the sense of being enveloped by family. Slowly, we began to envision how we could face the future. We were together, home in Wisconsin, on terra firma.

Meanwhile, the process of raising the geology department from the ashes continued. My new departmental colleagues and I realized that since we had to rebuild our program almost from scratch, we had a rare opportunity to create a geology curriculum no longer defined by the nineteenth-century emphasis on the systematic study of rock and mineral collections. Instead, we designed courses that reflected how the science is currently practiced, highlighting Earth processes and systems, with rocks and minerals as protagonists in larger plots. One challenge this posed was that there were no standard textbooks that mapped well onto the content of our new offerings. We each needed to develop new course materials, sometimes well outside our particular areas of academic specialization.

Two of my courses, one in high-temperature geochemistry and one on the long-term evolution of Earth systems, sent me deep into the literature on granite. I found that more than a half century after notions of granitization were finally abandoned, fundamental questions remain about how Earth became a planet with so much granite.

There are three major tectonic settings where granite (and its volcanic equivalent, rhyolite—often in the form of tuff) is forming on Earth today: continental arcs above subduction zones, such as the Cascades or Andes; continental rift zones like the East African rift; and continental hotspots like Yellowstone. Contra Bowen, none of these granitic melts come directly from the mantle. Instead, they are formed partial remelting of preexisting rocks when basaltic mantle magmas intrude various levels in the continental crust. Granites formed in each of these settings are subtly different in mineralogy and trace element chemistry, and in the bedrock of Wisconsin, we have examples of each.

Wisconsin's state rock is granite, specifically "red granite," quarried not far from our campus. An immensely popular building stone in the late nineteenth century—used, in fact, for Ulysses S. Grant's tomb—it is a classic-looking granite flecked with black hornblende, a water-rich mineral typical of continental arc settings. It was produced about 1.75 billion years ago, when Wisconsin was on the edge of North America and the seafloor of a forgotten ocean was subducting beneath the ancient Superior continent. Similar rocks are probably forming today beneath Mount Rainier and Mount Saint Helens.

Another generation of granites was emplaced beneath volcanoes in central Wisconsin around 1.45 billion years ago, forming a cluster of intrusions known collectively as the Wolf River Batholith. In the 1980s, the Batholith was briefly on the list for the nation's high-level nuclear waste repository but was eliminated, since most of the complex sits well below the water table and the risk of groundwater contamination was too great. The Wolf River granites are thought to have formed in a setting like modern-day Yellowstone, above a mantle plume where decompression melting of mantle rock led to secondary partial melting of rocks in the overlying continental crust. Wisconsin even has a few granites associated with the billion-year-old Midcontinent Rift, better known for its immense outpouring of basalt. These rift-related granites, like the Wolf River and Red Granites, have geochemical characteristics that point to remelting of older crustal rocks and not, as one might initially think, fractional crystallization of the magmas that yielded the Rift's prodigious volumes of basalt. Such crustal melting is also seen in the modern East African Rift.

All of this is reassuringly uniformitarian; we can move back and forth between the present and Wisconsin's Proterozoic past and recognize the same processes in action today. But upon further consideration, these sites of granite production point to a chicken-and-egg problem: for at least the past two billion years, it seems that granites have been generated mainly in places where continental crust, which is dominated by granitic rocks, already existed. Deepening the mystery is the fact that 70 percent of the continental crust on Earth today was formed prior to three billion years ago—before most geologists think the planet had even settled into its modern, subduction-dominated, granite-producing plate tectonic habits. That is, uniformitarian explanations can't account for the existence of most of the continental crust. Instead, we need to remove our Lyellian spectacles and consider what was different about the early Earth that might have allowed it to distill so much granite from its interior during its first billion years.

Compared with the middle-aged Earth of today, Earth in its youth had a much hotter mantle and higher rate of heat flow from its interior. The most direct evidence of this is an unusual volcanic rock type called "komatiite" (for the Komati River in South Africa), which stopped being erupted on Earth around 3 billion years ago, in Archean time. Komatiites were ancient lava flows with the olivine-rich composition of mantle peridotite and are, in a sense, "extinct." This is because today, even at the hottest volcanic "forges" like the midocean ridges and hotspots such as Hawaii, the temperature is not high enough for complete melting of mantle rock, and ordinary basalt is produced. Olivine, as Bowen showed, has an extremely high melting temperature and is left behind today as residual crystals at these sites (the famed green beach sands of Hawaii consist of grains of unmelted mantle olivine swept up with basaltic lavas and then weathered out of the rocks). Ancient komatiites tell us that at one time the mantle was hot enough for its full array of minerals—including olivine—to melt, thereby generating lavas of the same bulk composition.

The feverish thermal state of early Earth would have governed the character of its tectonic system. Geologists continue to debate about exactly when the planet began to practice modern-style plate tectonics, but most agree that subduction-driven tectonics, which requires ocean slabs that are cold enough to sink into the mantle and strong enough to pull whole plates with them, is unlikely to have occurred ab initio on a young, hot Earth.

And yet, the earliest crustal materials we can find are granitic. There is no actual rock record of Earth's first 500 million years, but a few tiny crystals of the tough mineral zircon have come down to us from that distant time, preserved as grains in an ancient sandstone in the Pilbara region of western Australia. Geochemical clues within those zircons indicate that they came from a granitic magma that had interacted with surface water—rain—4.4 billion years ago. The oldest surviving rocks are also granitic (though deformed and metamorphosed): the Acasta Gneisses of northwestern Canada formed 4.2 billion years ago, and they too bear evidence of interaction with water, as do slightly younger granitic rocks in the shield areas of all the modern continents. Collectively, these venerable old rocks hint at how Earth might have laid down the foundations for the continents before it ever adopted the custom of subduction.

One plausible scenario is an iterative bottom-up "refining" process involving many cycles of fractional melting facilitated by water.[9] A thin primary crust of basalt or komatiite— similar to the 4-billion-year-old highlands of Mars—may have been heated from below by vigorous mantle convection and partially melted in the presence of water to yield magmas with compositions slightly closer to granite. If that second-generation rock, in turn, partially melted, the magmas would be still more like granite, and so on. Such a process would have taken many iterations, but over the course of the planet's first billion ebullient years, a large volume of granite could be distilled from a very nongranitic mantle. A small planet like Mars that lost its internal heat early on, or a broiling one like Venus that boiled off its water, would never be able to get beyond the first basaltic step.

External factors may also have contributed to the young Earth's unsettled state. The face of the moon, with its dark, basalt-filled impact basins—Galileo's *maria* or "seas"—bears witness to a harrowing interval of meteorite strikes between 4.2 and 3.8 billion years ago that planetary scientists call the "Late Heavy Bombardment." Although there are no surviving impact craters older than 2 billion years on Earth—owing to the planet's remarkable ability to heal its wounds through erosion, sedimentation, and tectonics—there is no reason to think that Earth didn't experience the same fusillade. Some geologists have begun to speculate that the planet's prodigious early output of granite could be related to intense meteorite bombardment in its childhood.[10] On the moon, the giant impacts punched through the primordial crust and remobilized a rapidly cooling mantle, yielding the basalts that fill the lunar "seas." On Earth, an impact-shattered crust would have provided deep inroads for surface water that would have aided the process of partial melting and granite extraction.

The suggestion that one of the signature attributes of Earth—the granitic continental crust—had extraterrestrial origins is a radical proposal that challenges geologists' deep preference for Lyellian uniformitarianism. If the idea is correct, it implies that early cataclysms may have helped Earth to forge the stable, steadfast continents—and begin to keep a permanent record of its own story. I wish I could tell N about this bold new hypothesis; he would savor the audacity of it and immediately start pondering its implications. Of course! Why *wouldn't* the solid Earth have been changed by those ancient salvos? How extraordinary for the planet to use devastation as a chance to invent something new and durable. At least, that is my preferred interpretation; I need to believe that childhood traumas don't have to leave permanent scars but can instead lead to strength and resilience.

Once the sturdy interiors of all the modern continents were established, a cooler, more mature Earth settled into the habit of plate tectonics. From that point onward, the differences between granitic continental and basaltic ocean crust became more pronounced. Ocean crust has an entirely predictable life, like a child destined to carry on a generations-old family business. Its biography follows a simple arc from fiery birth at a midocean ridge to cold oblivion 150 million years later at a subduction zone. And like Tolstoy's happy families, all ocean crust is pretty much the same—remarkably homogeneous the world over—dutifully marching toward its preordained fate.

In contrast to the brief and predictable life story of ocean crust, the granitic continents have long, complex, idiosyncratic—not to say unhappy—biographies. Once created, granitic crust is not easily destroyed, except through the slow leak of turbiditic sediments onto the deep-sea floor. Continents are shuttled around the globe at the dictates of the single-minded ocean crust, always headed toward subduction, but they can expect to outlive any given ocean by billions of years, and endure multiple cycles of mountain building, rifting, burial, and exhumation, à la Hutton. And continents record all of these experiences, maintaining the most complete diaries on their partly submerged edges, the continental shelves.

The continental shelves are a liminal zone, simultaneously marine, because they lie below sea level, but also continental in the sense of being underlain by granitic rocks. Their hybrid character exempts them from the forces that destroy either fully continental or fully oceanic rocks. Because they are geologically continental and granitic, they cannot be subducted like ocean crust; but because they lie below sea level, they are not vulnerable to the ravages of erosion like most continental crust.

The continental shelves act as archivists for both the terrestrial and aquatic domains, taking notes about events on land that the eroded continents won't preserve, while also recording changes in ocean chemistry and marine ecosystems, which the short-lived ocean crust quickly forgets. Sediments of the continental shelves are our most detailed source of information about changes in the atmosphere, hydrosphere, and biosphere over geologic time, documenting evolutionary innovations, environmental upheavals, climate oscillations, and mass extinctions.

Still more remarkable is how these continental shelf archives are episodically reshelved by tectonics, like a library consolidating its volumes into a more compact space. When continents collide, the sedimentary logbooks of an entire era are pushed up into mountain belts, where they become accessible to land-based observers. In other words, an unlikely combination of factors ensures that chronicles of Earth's past are preserved over the long term—and available to any who may be interested. It's almost as if Earth *wants* its biography to be read, and granite is the substrate on which that story is written.

Given that granite, the basement rock, the foundation of continents, is the stage for human civilization, it's remarkable that we still don't completely understand its origins. Debates about granite began in the earliest days of geology and track the evolution of the discipline as a whole. Hutton's early intuitions proved correct, even though he flouted the rules of modern science. Old Werner the Neptunist wasn't entirely wrong—water is important for the formation of granites. Doris Reynolds and other granitizers, out of an honest desire to test prevailing paradigms, became hopelessly lost in a fiction of their own making. Bowen brought powerful experimental insights to the question but was unable to apply them plausibly to an old and complicated planet. Plate tectonics explained how granitic melts form today but doesn't resolve the question of how the earliest continents formed. Most recent, Lyell's foundational concept of uniformitarianism is being shaken by the proposal that the first large provinces of granite could be linked with primordial meteorite impacts. Granites remind us that rigid orthodoxies have no place in Earth science. Even as we have come to know it better, the planet defies our most sophisticated models and remains deeply mysterious. But the ground beneath us is no less solid for holding secrets.

Three years after moving the geology department from its old quarters, we relocated again to a gleaming new space in a renovated building. This time, I had two colleagues to help, and there were no hidden subbasements to discover. All our specimens would now reside in well-cataloged cabinets—including several filled entirely with Wisconsin granites—and for the first time in its history, the department was above ground, on the sunny second floor.

The children and I were also emerging into the sunlight from the dark tunnel of grief. My parents and I strove to create a warm, well-lit sanctuary space for the boys, assurance that they were safe. My father built a tree house in our yard with one floor for each of the three. On Saturday mornings, they would race down the block in their pajamas for grandma's cinnamon rolls. In winter, they trundled off in snowsuits to the sledding hill behind my parents' house. Some of their teachers became surrogate aunts and uncles, and geology students served as indulgent big brothers and sisters. Little by little, we reshaped ourselves into a new, fierce, postnuclear family. Emerging from the rubble, we were grateful to feel the good granitic earth beneath our feet again.

ECLOGITE

It's a luminous summer night on a rocky, wind-scoured island in western Norway. The boys should be in bed—it's after 11:00 p.m.—but the magical feeling of the place has bewitched all of us. We're an odd assortment of people, brought together by rocks: an eminent Norwegian geologist; his shy teenage daughter, who is helping to look after my kids; two graduate students, one Swedish and one German; and me. We're playing a precarious game of catch just outside the cabin, adults and children scattered across an archipelago of rocky knobs separated by seas of *lyng*, or heather. Every toss is a risk—if the ball falls into the dense mat of vegetation, it will be lost. But tonight we are superhuman, and each successful catch is greeted with jubilant cheers. At last, the sun slides into the sea, and the kids, happily exhausted, go to sleep without protest.

We adults talk for a while then climb into cupboard-like bunks typical of Norwegian mountain *hytter*, cabins. I realize that I have an unaccustomed feeling of well-being—that I am home, *hjemme*, as if an atavistic memory that has survived two generations in America has been rekindled. Snug under the duvet, savoring the coolness of the air, I feel a wave of something like grace wash over me. The boys and I have left behind the horrors of illness and death, and now we're here, safe, in a pure, elemental place with nothing but rocks, sea, sky, and *lyng*. Tonight, I think, we've experienced what the philosopher Mircea Eliade called *hierophany*—literally, "the sacred made visible"—the divine breaking through into the everyday world.

With geologists from the University of Oslo, I'm here to study the extraordinary rocks on this rugged island, which provide a rare glimpse into the hidden process of metamorphism. In the dazed months after N died, I felt like a high-functioning zombie. My long-practiced habits of time management and hyperorganization allowed me to keep up with the practical demands of caring for the kids, maintaining the household, and teaching my classes, but emotionally, I was incapacitated. I had even lost my appetite for reading, the one thing that had always brought relief when reality became too oppressive. The magnitude of what had happened to us made the plots of novels seem flimsy and self-indulgent. I needed a new storyline for my own life.

I had a sabbatical coming and began to imagine a year of geological therapy. I had read scientific articles about the unusual rocks on the island and realized I had an indirect connection through Ohta to the geologist, H, who had first recognized their significance. The two of them had spent a season together in Antarctica and held each other in high esteem. I wrote to H about the possibility of collaboration, and he welcomed me to join his group the following year. Days before the deadline, I applied for a Fulbright fellowship, and to my astonishment, was granted one. It provided just enough for the kids and me to rent a flat on the edge of Oslo.

We arrived in July to get fieldwork done before university classes began and to give my kids some time to acculturate and learn some Norwegian before being plunged into school. H looks a bit like my dad and his brothers. In fact, almost everyone we encounter on the western islands reminds me of relatives in my parents' overwhelmingly—some might say oppressively—Norwegian hometown in northern Minnesota; it seems like there are about a dozen types of Scandinavian faces that just keep getting allocated to new generations.

H was born on a nearby island and had been expected to become a fisherman like his brothers, father, grandfather, and countless earlier generations before them. But, to the great shame of his family, H suffered terribly from seasickness and preferred to stay on solid ground. Staying on the shore while his brothers were at sea, he became interested in rocks and eventually earned a PhD in geology. Despite his status as a professor and scientist with an international reputation, he says his family still considers him a failure. I suspect he is being much too modest.

One evening, H takes my boys down to the shore and shows them two marvels: a Viking grave site, and also a place where they can peel large flakes of pearly mica from the rocks. Seeing them all bent down together, absorbed in the mystery of these things, I am hit with a visceral sense of what the boys will be missing without a father in their lives, and a sob rises in my throat.

There is a special kind of intimacy that comes from the shared experience of wonder, and this island is a place where there is so much to wonder about. The rocks here are in some ways cousins to the Midcontinent Rift rocks in Minnesota—not only about the same age but also mostly basaltic, and formed in an ancient rift setting. But unlike their North American counterparts, which are tilted but otherwise not much changed, the Norwegian rocks have been radically altered by two distinct episodes of tectonic upheaval that took them deep beneath the surface. They are chthonic rocks, shaped by their experiences in the underworld.

Their first tectonic transformation happened around 900 million years ago, during continental collisions that formed an ancient supercontinent called Rodinia, a generation older than Pangaea. The rocks in this part of Norway were deformed and heated to extreme temperatures (greater than 1,500°F) in the heart of the resulting mountain belt, which converted them into an unusually dry metamorphic rock called granulite. On Earth, so awash with water, granulite is peculiar. It contains no hydrous minerals like flaky micas or black hornblende, common to many igneous and metamorphic rocks. And because of its extreme state of dehydration, it is exceptionally strong. When water gets into crystal lattices, even in trace amounts, it makes rocks much weaker than they otherwise would be. Dry granulite can tolerate stresses that would cause ordinary water-bearing rocks to fracture or flow. About two-thirds of our enchanting island, including the site of the cabin and the pedestals where we played catch, is underlain by granulite.

The rest of the island is made of an even more distinctive metamorphic rock: eclogite. It is so rare at the surface of Earth that many geologists have never seen it in outcrop. But eclogite is arguably one of the most important rocks on the planet because it is the motive force behind Earth's tectonic system. Everything about eclogite is exceptional. Visually, it is an exquisite jewel-box of a rock, with orbs of red garnet in a matrix of grass-green sodium pyroxene and blades of blue kyanite. These polychromatic minerals form when black basalt is subjected to very high pressures but comparatively low temperatures—an unusual combination of conditions that occurs almost exclusively in subduction zones.

The name eclogite comes from the Greek *ekloge*, to select out; a term used in theology for those whom God has chosen for special purposes. This seems apt, because eclogite has a very special role to play in Earth's active-lid convection system. The conversion of basalt to much denser, garnet-rich eclogite is the reason that subducting slabs of ocean crust are able to sink deep into the mantle. Without this transformation, plate tectonics as we know it would not function. Ocean crust would pile up at subduction zones like cars rear-ending one another in stalled traffic; water would not be returned to Earth's deep interior, and the viscosity of mantle rocks would likely be too high for convection to persist.

One of the strangest and most wonderful things about Earth is that basaltic ocean crust—produced by partial melting of mantle rock—can be recrystallized at high pressure into another form, eclogite, that is denser than the mantle. This process of conversion perpetuates the convective cycle that keeps the exterior and interior of the planet in communication, endlessly renewing the atmosphere and hydrosphere and providing the raw materials for life. I often say a prayer of gratitude for eclogite, the chosen rock, for keeping the planet running.

On our island, the eclogite formed in a second cycle of supercontinent formation and mountain building, during the rise of the Caledonian-Appalachian chain and the assembly of Pangaea about 400 million years ago. I've explored other parts of that ancient mountain belt, in Svalbard, Ellesmere Island, Scotland, and West Virginia, but these rocks take me farther into its interior than ever before. During the Caledonian deformation, the dry granulites, which had cooled since

the earlier tectonic episode, found themselves at the bottom
of a stack of rocks deep inside the new mountains, more than
thirty miles below the surface. Under such pressure, they
should have converted to garnet-rich eclogite. But in this
place, that conversion was incomplete: patches of rock—now
the low-lying areas of *lyng*—did become eclogite, but much
of the older granulite—our pedestals—persisted, unchanged.
For some reason the process of metamorphism stopped mid-
way through, and we're trying to understand why.

We suspect that a third rare rock type that occurs only as
thin black veins in the rocks may be the key: pseudotachylyte.
Pseudotachylyte is a dark, glassy rock formed when frictional
melting occurs along a fault during an earthquake. Because
the ambient rocks are much colder, the melt then quenches
rapidly—probably in minutes—to a noncrystalline solid, or
glass. Other types of fault rocks can occur through slow, non-
seismic slip, but pseudotachylyte is considered an unambiguous record of an ancient earthquake. Even many geologists
find its name befuddling because they've never heard of "true"
tachylyte—an uncommon type of obsidian—much less an er-
satz version. Despite its obscure name, however, pseudotachylyte rewards study because it provides a window into what
happens on a fault zone far down in the subsurface at the time
of an earthquake. These processes can't be observed in real
time; exhumed pseudotachylytes are the closest we can get to
witnessing the deep origins of earthquakes.

H introduces me to outcrops where the geometric relationships among the rock types are clear, and I make detailed
notes about the "habitats" of the pseudotachylyte. We identify
spots where specimens for microscopic examination would be
most revealing. The rocks are too hard and glacially smoothed

to be sampled even with a sledgehammer; we must employ a noisy rock drill that requires a steady flow of water to lubricate the bit. Up in the area where the rocks are best exposed there aren't any natural sources of fresh water, so we have to carry it in tanks from the cabin. H and his students are focusing on the mineral assemblages in the rocks to better constrain the pressures and temperatures they experienced on their travels down and up and down and up again. My job is to bring a geophysical eye to these strange visitors from a hidden world. I've never seen rocks that have made such epic journeys. *Hierophany*, I think again.

Summer comes, we settle into life in Oslo, and my kids plunge bravely into Norwegian schools. O is in the third grade and F, in first. K is in *barnehage*, or preschool. I can see how hard it must be for them, and I feel pangs of guilt for dragging them across an ocean for my self-prescribed rehabilitation. But I also want them to understand their ancestral roots— and more generally, to expose them to the world beyond our midwestern town. My heart swells as each finds a way into their new cultural settings. For O, reticent and shy, Pokémon cards are the lingua franca that help him become friends with classmates. I would never have expected to feel gratitude for the likes of Charizard and Bulbasaur. Rough-and-tumble F is in his element in grade one, in which Norwegian children spend most of their time outside, hiking or skiing, building forts, cooking on campfires, and learning how to be members of a cooperative society. K, not quite three, is at the optimal moment for language acquisition and soon speaks more fluent Norwegian than I do.

I have been given an office with a skylight in the top floor of the elegant old Geological Museum in the district of Oslo called Tøyen, on the grounds of the university's nineteenth-century botanical gardens. H is apologetic that there isn't space for me in the modern building on the main campus, but I am grateful to be in this serene space, where I don't need to perform the role of being a normal person. I may look like an ordinary woman in her late thirties, but I feel broken by the events of the previous years. The rarefied space of the museum, with its orderly display cases and overstuffed library, is exactly where I want to be. Depending on the time of day and the way light comes through the windows, the building exudes whiffs of Belle Epoch aspirations, early twentieth-century Norwegian nationalism, and wartime austerity. My life has become more complicated than I could have imagined, veering far off the main road. I am comforted by the sense of a collective past that pervades the museum, a throughline across the decades. In good times and bad, the museum has never doubted its purpose.

My office is next door to one once occupied by a towering figure in the geosciences, Victor Goldschmidt. Born in Switzerland in 1888 to a Jewish family that emigrated to Norway when he was thirteen, Goldschmidt considered himself a loyal Norwegian. After earning his PhD at the University of Oslo, he established the Statens Råstofflaboratorium—the National Raw Materials Laboratory—to inventory Norway's mineral resources, which became essential for strategic purposes in both world wars. In 1929, Goldschmidt took a prestigious faculty position at Göttingen University in Germany but was forced to leave in 1935 as antisemitism tightened its grip. He fled back to Oslo and then narrowly escaped deportation to a concentration camp when Norway was invaded by the Nazis in 1940.[1]

Goldschmidt articulated many of the foundational principles in the study of igneous and metamorphic rocks—including meteorites—and was nominated eleven times for the Nobel Prize in chemistry. In 1988, in belated recognition of his contributions, an annual international conference in geochemistry was named in his honor.[2] Goldschmidt was an acquaintance of Einstein, and in the hallway outside my office, there are photos of them in the wooded plateau above Oslo called Nordmarka sometime in the 1920s. It's a static, slightly stilted shot, but nonetheless I can sense the energy of that day, the swaying trees, the restless air, the electric conversation interrupted by the photographer. Two decades later, the dense forests of Nordmarka hosted clandestine camps for Norwegian resistance fighters.

While the children are in school, I immerse myself in the logic of the rocks on the island, in the geological equivalent of scriptural exegesis. Alternating among the geologic map, my field notes and photographs, and thin sections under the microscope, I try to re-create the sequence of events encoded in the rocks: to see them cinematically. The starting point was the dry granulite, and the end was the messy partly eclogitized mass we see today, but what happened in between? The pseudotachylytes—records of massive ancient earthquakes—must be central to the story, but how?

H has shown that the pseudotachylytes contain tiny crystals of the classic triad of eclogite minerals: garnet, sodium pyroxene, and kyanite. Their lacy shapes indicate they grew quickly from a melt, not by later metamorphic recrystallization. This means that the earthquakes that formed the pseudotachylyte were unusually deep—occurring thirty miles or more below the surface. In the continental crust,

most earthquakes happen at shallow depths, where rocks are
strong and brittle; counterintuitively, only strong rocks can
generate earthquakes—they must be able to store stress to
release it seismically. Weak rocks, in contrast, simply ooze or
flow, never building up enough stress to fail in an earthquake.
Normally, the change in mechanical behavior between strong,
earthquake-producing rocks and weak, oozing rocks occurs at
a depth of about ten miles, where temperatures become high
enough that rocks begin to flow in the solid state. Geologists
call this the "brittle-ductile" or "brittle-plastic" transition. The
fact that we are seeing evidence for earthquakes more than
twenty miles below that level means that the rocks on our
island were anomalously strong.

Studying the distribution of pseudotachylytes on the
map, I realize that they occur only in the granulites, not in
the eclogites, even though we know they formed at eclogite depths. This is a hint that they probably happened early
in the process, and that the eclogites were too weak to fail
seismically. The latter inference is consistent with H's observation that the eclogites have small amounts of water-bearing
minerals—including the mica he showed my boys down on
the shore. H has surmised that water was critical to the formation of the eclogites; without it, the process of metamorphic
recrystallization of the granulite was simply too slow, even
though the granulites were many miles below the depths at
which eclogite formation should have begun. Without water,
atoms can only move through rocks by the arduous process of
diffusion, which is tortuous and slow, like driving in gridlock.
In the presence of water, atoms can slip into solution and
hitch a ride, like hopping on the subway and moving quickly
through the city.

I look up at the clock and realize with panic that I have only thirty minutes to get across Oslo to pick up K from *barnehage* and O and F from their after-school program. Fortunately, public transportation in the city is excellent; I leave the museum, dash to the metro station, and gather the three kids from two different sites with five minutes to spare. We don't have a car here and don't need one. In Oslo, the metro (T-bane), electric street cars (*trikken*), buses, and ferries are integrated into a seamless network, and we easily find our way all around the city. One T-bane line ends in the vast Nordmarka forest. In the rocky uplands, where resistance fighters once hid from Nazis, there are now miles of bike and ski trails—and on Sundays, itinerant vendors selling warm waffles with strawberry jam. With no teaching obligations, I am able to enjoy weekends with the kids without the coming week's lectures and assignments looming over my head. I feel lucky for that—and, when Monday comes, for the privilege of returning to the world of our half-metamorphosed rocks.

Gradually, I began to have a sense for how the granulite-pseudotachylyte-eclogite story might have unfolded. During the building of the Scandinavian Caledonides, the old, dry granulites were overthrust by other rocks until they were deep in the roots of the mountain belt. Surrounding rocks may have been changing to denser forms, but the ultradry rocks remained intact until finally one or more large earthquakes broke them open, allowing deep waters to make inroads into the rocks and triggering their long-suppressed recrystallization to eclogite.

At first, this process would have been self-perpetuating—a positive feedback—because the conversion to eclogite would have made the rocks denser and more compact. This shrinkage would have created tiny tensile microfractures at the boundary between the eclogitized and the unconverted rock. These fractures, in turn, would have allowed water to infiltrate farther into the rocks, leading to more eclogite formation, and so on, until the water had been consumed in the formation of micas and other hydrous minerals. In fact, in outcrops on the island we could see "fronts" where the eclogite process had abruptly halted, probably because the water ran out.

But why didn't water enter the entire complex and complete the conversion? I wondered if there might be significance to the fact that across many spatial scales, from the island as a whole to the field of view under a microscope, only about 30 percent of the rock mass had converted to eclogite, while 70 percent persisted as granulite. It seemed that for some reason, the eclogite-forming process was arrested after about only one-third of the rock had undergone the transformation. This reminded me of studies I'd read on the physics of magma, which showed that when rock is heated to the point of melting, it changes abruptly from a rigid solid to a viscous fluid when just 30 percent becomes molten. That is, the change in bulk strength doesn't happen in a steady, linear way as the amount of melt increases. Instead, there is a threshold value—what igneous petrologists call the "rheologically critical melt fraction"—where the crystalline part no longer forms a stress-supporting scaffolding and the entire mass suddenly becomes as weak as the melt. ("Rheology" comes from a Greek root meaning "to flow.") A similarly abrupt transition in bulk rock strength could be the key to what stopped the metamorphic conversion of the granulite to eclogite on our island.

Here is what the rocks seemed to be telling us. First, one

or more large earthquakes violently fractured the dry, strong granulites, admitted water into the complex, and began to convert the rock to eclogite. Initially, eclogite formation spawned more eclogite through the creation of microfractures. But the process was ultimately self-limiting because once the volume of eclogite—water-bearing and much weaker than granulite—reached the critical fraction of around 30 percent, the rock mass behaved as a viscous fluid, no longer capable of storing stress. And if it couldn't store stress, it would never fail seismically. With no more great earthquakes, there was no further infiltration of water, and thus the rocks had remained in their strange, half-converted state, allowing us this rare, *hierophanous* glimpse of the process of metamorphism in action.

Our odd rocks were conveying an essential truth about Earth: metamorphism—so central to the rock cycle and tectonic system—is entirely dependent on the presence of water in the crust. In the absence of water, metamorphic reactions are simply too sluggish to allow rocks to equilibrate to their surroundings. Like tourists who travel to new places but do not actually engage with other cultures, dry rocks could find themselves at new pressures and temperatures but remain stubbornly unresponsive. Without water, subduction, which requires the metamorphic conversion of basalt to eclogite, would not happen, and Earth's elegant, self-renewing tectonic system would not be possible.

Because water is so ubiquitous on Earth, it is easy to overlook its central role in virtually every geologic process. The rocks from our Norwegian island, with their unusual tectonic history, provided a kind of counterfactual view of Earth as it might have been without so much water in the crust. Although the two-stage metamorphic history of our rocks seems rather unusual today, it might not have been in Earth's distant past, when the planet was hotter and dry granulites would have been formed at shallower depths. Our work suggested that such rocks would never fully convert to eclogite—further support for the idea that modern-style plate tectonics would have been unlikely to operate on the early Earth.

Gleaning these insights from the rocks was deeply satisfying, an almost spiritual epiphany. These strange rocks had spoken to me across deep time. I felt my joie de vivre returning and happily set about writing up our findings for publication. As I described our model for how the metamorphic process had stopped itself, I became curious about the physics behind the concept of the rheologically critical fraction. Scouring the scientific literature for other examples of this phenomenon, I found that it had an astonishingly wide range of applications, from cheese making to blood clotting. Einstein himself had even written a paper on the behavior of rigid spheres in a fluid and how the effective viscosity of such a mixture changed in a nonlinear manner.[3] I savored such serendipitous discoveries—the feeling that something about the underlying logic of the world had been revealed.

Living in Norway brought other revelations. In our Norwegian American family, *Norge* was always spoken of in reverential terms as the motherland, the source of all that was true and good, a hallowed, heavenly place (even though it had not been benevolent enough to support our immigrant ancestors). A family story that left a deep impression on me was of a great-great-aunt on my father's side who for years after coming to the flat prairies of northwestern Minnesota would sit and weep for the beautiful, unfarmable mountains back in Norway.

By the time of my sabbatical at the University of Oslo, I had visited Norway many times in transit to and from Svalbard, and also to visit distant cousins and hike in the mountains. I had taken Norwegian language classes in graduate school. I knew the work of Henrik Ibsen and Knut Hamsun, Edvard Munch and Edvard Grieg, Sigrid Undset and Liv Ullman. But this was my first experience of being in Norway with children, and of staying through the *mørketid*, the dark wintertime. I gained a better understanding of the roots of the Scandinavian commitment to collectivism: people had to look after each other to endure the long, sunless winters.

There are so many good things about Scandinavian social democracy, especially for families with young children—not only excellent public transport but also universal health insurance, affordable childcare, dental visits in schools, ubiquitous walking and biking paths, and a general sense of well-being, that there is plenty to go around and no one needs to fear slipping into destitution. On the other hand, everything is expensive in Norway, and the Fulbright stipend barely covered the cost of living. But there were so many things the children and I could do for free: watch boats coming and going in the harbor; ice-skate on a rink just outside the Parliament building; visit playgrounds with Viking ships and zip lines; pick berries and skip stones from the shores of a lake near our flat.

We lived on the western edge of the city, near an old, prosperous eighteenth-century farmstead that had been converted to an agricultural museum. One day in October, my middle son's first-grade class visited the farm to help with harvesting root vegetables and he came home staggering under a backpack of potatoes that weighed as much as he did. The potatoes lasted for months, and some of the accompanying soil—(earth from the motherland is not "dirt")—deep in the seams of his pack, came back to Wisconsin with us. I learned from a brochure that was sent home with the children that the farm had belonged to the eminent Wedel Jarlsberg family—for whom my PhD area in Svalbard was named. It was a small connection that made it seem that we belonged in this place.

It was a pleasure to be able to say our last name (which even some of my close friends find difficult) to bank tellers or receptionists and have it understood immediately, with no need for spelling and respelling—and written with the "ø" restored to its proper place. Many other aspects of life in Norway also felt right and familiar to me. Norway's dark winters and austere landscape make it impossible to deny the authority of Nature. The formidable terrain instills a kind of resignation—not defeatism, but an acceptance of limitations imposed by reality. The precipitous walls of the fjords remind you that existence is precarious, and life is not supposed to be easy. Just be grateful for a warm bed and a lamp on a January night. The folk music of Norway, centered around the haunting open fifths of the Hardanger fiddle, expresses this emotional state: neither happy nor sad, or perhaps, both happy and sad. Beauty and meaning, the fiddle sings, are inexorably entangled with struggle. This psychological set point seems healthier than the American pursuit of unalloyed happiness, which, ironically, engenders perpetual dissatisfaction (and fuels the engines of capitalism).

Another foundational Norwegian principle, *allemannsretten*—the freedom to roam on private land—was already something I often practiced, furtively, in the United States. ("Rocks belong to everyone" has always been my planned defense to charges of trespassing.) And there is the unwritten law of *janteloven*, the consensus that "tall poppies" should be lopped off, that personal ambition is bad for society. I certainly internalized this as a child—to my detriment as an adult in the ego-driven realm of academe. Taken to its extreme, *janteloven* can lead to stifling conformity, but in a world where social media has made constant self-promotion seem normal, a little *janteloven* would be a refreshing corrective.

I suspect that *janteloven*, like other Norwegian norms, emerges from the landscape itself. Nature is clearly in charge; the land is rocky and the seas are rough. In the face of such conditions, humility is the logical response. In mainland Norway, place-names—many surviving from Viking times— reflect this; very few cities or landscape features are named for people other than a few kings and old gods. Most are simple, pragmatic topographic descriptions. Some examples en route from Oslo to our partly eclogitized island include Jevnaker (flat acres); Myrdal (swamp valley); Bergen (the mountain). The simple declarative nature of these names stands in contrast to the pretentions, euphemistic, and history-erasing eponyms all over the US map. In Ohio alone, consider Oxford, Athens, Defiance, Napoleon, Columbus. The names of American suburban streets evoke idyllic natural settings—Forest Glen Court, Prairie Ridge Road—but these are typically memorials to ecosystems destroyed in their development. Over time, I believe, routine exposure to such half-truths corrodes the soul.

In many ways, I felt culturally more at home in Norway than I ever had in the United States; some core Norwegianness had survived two generations in our family. But to Norwegians, I was thoroughly American. It was sometimes frustrating to begin speaking Norwegian and have the conversation shift immediately to English. Norwegians are far less accustomed to hearing their language spoken by foreigners than speakers of English are. In the United States, in particular, we all learn early on to listen flexibly to a wide range of accents. And I sometimes felt frustrated with Norwegian naivete about multiculturalism; in a small homogeneous country, it's easy to think everyone can just get along. Yet Oslo was a notably segregated city, with its immigrant population living almost entirely in Tøyen, the district with the Geological Museum and Botanical Gardens. A group of Romani people camped for a month in a parking lot across from the museum, and I cringed at what I overheard Norwegians saying about them in a nearby café.

I also learned later that Victor Goldschmidt himself, whose vast knowledge of Norway's rocks and minerals had been so essential to the nation's security and economic growth, faced anti-Jewish prejudice in Norway after World War II.[4] Goldschmidt had fled for his life from occupied Norway to neutral Sweden, then England, in 1942. When the war ended, he had hoped to resume his post at the University of Oslo, but his former colleagues blocked his reappointment, based on a false rumor that Goldschmidt had accused another member of the faculty of supporting the Nazi-sympathizing Quisling government. Goldschmidt nonetheless returned to Norway in 1946, then died suddenly of a cerebral hemorrhage in 1947. Before the war, he had made stone crematory urns from distinctive Norwegian rocks for his parents and himself, and for decades, his ashes were kept in one of these, a green peridotite vessel, in a cupboard at the Geological Museum. He was finally interred a half century after his death at an unmarked grave in Vestre Gravlund, Oslo's largest cemetery. His snubbing was almost certainly motivated by antisemitism, but his international renown was also an unforgiveable violation of *janteloven.*

Still, Norway plays an undisputed global role as a beacon of idealism and defender of human rights. Alfred Nobel, the Swedish inventor of dynamite who established the Nobel Prizes, stipulated that the scientific prizes be awarded by the Royal Swedish Academy but that the Peace Prize winners be selected and honored by the Norwegian Parliament, reputedly because he trusted Swedish academic standards but not its militaristic tendencies. At the time of Nobel's death in 1896, Norway was actually still under Swedish rule; it did not become a fully autonomous country until 1905, and thus had no army of its own. But perhaps Nobel also recognized the fundamentally peaceable, sweetly naive nature of Norwegians (at least in the post-Viking era). Even after independence, Norway did not invest much in a standing army, and the country was not prepared when the Nazis arrived in April 1940. Today, military service is compulsory for all young men and women in Norway, and a former Norwegian prime minister (Jens Stoltenberg, whose last name means "proud mountain") has had a long term as Secretary General of NATO. But the Nobel Peace Prize is still chosen by a Norwegian committee and awarded in Oslo.

Each year, the Norwegian Fulbright office receives one ticket for the Peace Prize ceremony and puts the names of all the visiting scholars in a hat to determine who will be able to attend. One late autumn morning, I got the thrilling news that my name had been chosen. Within a few hours, my exhilaration turned to panic as it dawned on me that my closet of thrift-store apparel contained nothing appropriate for such an event. I spent an exhausting day in upscale stores on Karl

Johans Gate, Oslo's main street, feeling painfully out of place among the chic clerks and customers. I finally found a simple but elegant dress I could afford, shoes that could pass for stylish, and a small, discreet purse to carry. It didn't occur to me until the day before the event, however, that my puffy winter jacket would not be acceptable outerwear. Since it was only a few steps from the T-Bane station to the Rådhus, or city hall, building where the ceremony is held, I decided to forgo a coat altogether. Luckily, it was a rather warm December afternoon.

To my relief, my outfit did not seem to violate Nobel fashion protocols, and I was ushered to a seat next to a well-known British actress, who had no obvious connection to the honoree, South Korean president Kim Dae-jung (but then again, neither did I). The acoustics were quite terrible in the hall, with amplified voices bouncing off floors and walls of marble from southern Norway. But I felt so lucky to be there, in the presence of diplomats, heads of state, and the king of Norway himself. I wished that I could tell my homesick great-great-aunt: "I made it back to the old country. I met the king."

The poor, overpopulated, agrarian Norway of my immigrant ancestors is rapidly fading from memory. Norway had even undergone a partial metamorphosis between the time I started working in Svalbard and the year we lived in Oslo. I could sense that a changing of the guard was underway, a transition of power from the generation shaped by wartime occupation and postwar austerity to one that had known Norway only as an affluent oil-producing nation. During my first season in Svalbard, the hiring of a young German geologist at the Norwegian Polar Institute had caused great controversy. Senior scientists who had served in the Resistance made impassioned pleas not to hire the young man, even though he had been born a dozen years after the war ended. Now, almost two decades later, he was an esteemed long-term member of the staff. More broadly, the frugal habits of the war years in Norway had given way to more American-style consumerism, and the huge national nest egg of oil money (*Oljefondet*, The Oil Fund) for future Norwegians has fundamentally changed the collective psyche. The country is rich now; its former poverty is now remembered only by the oldest citizens.

A recent study suggests that the concept of the rheologically critical fraction, which was the key to understanding our partly eclogitized granulites, may apply to human societies.[5] Controlled experiments on consensus-building, together with mathematical models of human networks, show that the scaffolding of shared beliefs that unites a community can be abruptly destabilized by a vocal dissenting group representing only 25 to 30 percent of the population. That is, a small but well-networked minority can radically change the way the whole society functions.

This "sociologically critical fraction" behavior is not inherently good or bad. Depending on the demographics of a given country, it governs when members of a given generation will take the reins of cultural power from their elders. The existence of such a low critical fraction is heartening if an abhorrent system like racial apartheid is the prevailing view, because it suggests that enlightened change can occur rapidly. But the concept also has a dark side: A seemingly strong democracy can be undermined by a minority that disseminates corrosive falsehoods insistently enough. Seeing evidence for this phenomenon in the rock record makes such a scenario all too plausible to me.

After a year in the underworld with granulite and eclogite,

I felt ready to return to the surface and rejoin the hubbub of everyday life. For me, metamorphic rocks are the wisest, most interesting mentors in the pantheon of stones. Seasoned travelers, they are exemplars of versatility, illustrating how immersion in unfamiliar and uncomfortable surroundings can lead to transfiguration. In most instances, metamorphic rocks record only their final destinations: milky marbles have mostly forgotten their marine origins as limestone; garnet schists are only dimly aware of their beginnings as lowly mudstone. But the partly eclogitized granulites on our enchanted island hover forever in a liminal state of being neither and both, on a threshold between worlds, like our immigrant ancestors who lived in America but clung to Norway. The boys and I had crossed that threshold in the other direction. The land itself—its deepest bedrock—had brought us back to our origins, where nature still looms large. Coming through customs on the way back to the United States, our American passports now seemed inaccurate: we were citizens of Earth. We were home and ready to settle into the ordinary business of being Earthlings.

9

GLASS AND FLINT

It's the middle of an August night in the Marche region of Italy, in a tiny village whose buildings have been repurposed as a geologic field school. I'm asleep in the stone house reserved for faculty. My three sons, now all in their twenties, happen to be visiting for the week and are in an old villa next door. A strange rumble, like a deep exhalation from Earth itself, shakes me from dreams into consciousness. I look at the clock: 3:36 a.m. A sudden wind in the trees, perhaps. But then my brain snaps to full alertness and I somehow know that this is the P-wave—the first subsonic pressure pulse—of an earthquake. A window breaks, a wardrobe teeters, and I rush to the door before the more damaging roller-coaster S-waves arrive. Dogs begin to howl, roosters crow, and wild boars come rushing out of the woods. In the morning we learn that three hundred people have died in the nearby town of Amatrice. This tranquil, beautiful place suddenly feels feral and malign; I've put all of my children in mortal danger. My intellectual understanding of seismicity did not prepare me for how awful it is to feel solid rock convulse.

My sons and I are accustomed to a background level of tumult; as the three of them rolled through elementary, middle, and high school, I pictured our tough little family as a stone tumbling in a rushing stream. We hurtled onward, mostly too dizzy to realize that time was passing or where we were going. The combination of single parenthood and my "day job" as a professor required all the logistical expertise I had gained in years of Arctic fieldwork, and there was some joy in the high-wire act of keeping everything in motion. Other times it felt as if I were pulling a heavy sledge behind me, like an old polar explorer, while peers glided effortlessly by.

We were constantly busy and surrounded by people, but our family didn't fit easily into faculty social circles, which tended to be centered either around couples or single people without children. Craving some semblance of adult life, I had a series of ill-judged boyfriends who, in the end, only reaffirmed to the kids and me how far outside the mainstream we were. Having made our way together through a dark wood (sensu Dante), we were averse to the bright lights of popular culture. We were blasé about football in a place where the Green Bay Packers are demigods. We were vegetarians in a land of burgers and bratwurst. Out of frugality and environmental principle, I got most of our clothes from thrift shops or rummage sales, and dried laundry outside, summer and winter, on a line looped around the apple tree. We had a rain barrel, compost pile, and vegetable garden in our small backyard. We foraged around town for wild berries and grapes. These small acts of rebellion against the juggernaut of consumerism helped me maintain some sense of dignity and autonomy.

Throughout these years, my parents were our bedrock, a constant, reassuring presence, and I was thankful that the children were growing up in a multigenerational embrace. My father spent hours making projects with the boys in his woodshop, and my mother shared with them her love of books. My parents generously looked after the kids so I could attend professional meetings and even return to Svalbard for two abbreviated field seasons. I realized that there were some advantages to being a one-adult household—especially the relative ease with which the boys and I could relocate temporarily as academic opportunities arose. Although Wisconsin was our cozy base camp, and the boys often came with me on geology department field trips, I wanted to share adventures with them in other parts of the world. So, a few years after our stay in Norway, I moved the family temporarily to London, where I taught in our university's off-campus program, and then later to Dunedin, New Zealand, for a sabbatical. Each time, the kids heroically entered new schools and learned to assimilate.

If parenting were only about logistics, it would be comparatively easy. When the boys were small, and their needs were simple, I believed that I could shield them from harm with my maternal superpower—an impenetrable force field of ferocious mother-bear love. As they grew older and the outside world drove each into different paths through the labyrinth of adolescence, I felt less and less qualified as a parent. How could I possibly prepare them to enter the aggressive, unforgiving realm of adult men? In addition to the usual sorrows and struggles, illnesses and mishaps of everyday life, there were larger, intractable events that caromed us into unlit corners: for my sons, the suicide of a close friend; for me, my sister's ongoing travails and the premature loss of childhood companions, including my "city mouse" counterpart, whose father's Minneapolis furniture store had once suggested so many possible futures. Her overdose death was an unspeakable tragedy. All of us shared the terror of my mother's encroaching blindness and near-fatal brain tumor. For a time, it seemed that the aftershocks of one event had barely subsided before the next one knocked us off-balance.

But we had come through all of that. My mother had rebounded after a nightmarish fortnight in the ICU; my sister had come back from California, become an enrolled member of her birth mother's Ojibwe tribe in northern Wisconsin, and had established a stable life for herself. All three of my sons had attended good colleges and were beginning to chart their individual life paths.

And now we were in the Italian Apennines together, able to enjoy each other's company as adults. It had become rarer for the four of us to be in the same place for more than a day or two at a time. We were savoring the reunion in this sublime and rocky place when the earthquake jarred us out of our feeling of fragile equilibrium. I felt somehow guilty, revealed as a hypocrite, for having studied earthquakes with such academic enthusiasm. Earthquakes are horrible.

Most geologic processes are famously slow, but in an earthquake, whole mountains lurch. In the magnitude 6 earthquake we experienced, a slab of rock about eight miles long and six miles deep slid a foot in under a second. In the largest earthquakes—magnitude 9 events at subduction zones—slip can exceed fifty yards on surfaces hundreds of miles long and tens of miles deep. After my work in western Norway on rocks shattered by ancient seismic events, I had developed an almost obsessive interest in pseudotachylytes, the glassy "fossil earthquakes" that record frictional melting along faults. After years of investigating geologic structures that had been formed in extreme slow motion, over time scales too long to be witnessed by humans, I found something irresistible about rocks recording cataclysms that had occurred in "real time." I felt something like empathy for them and wanted to see what traces such violence leaves.

And so, eight years before we experienced the earthquake in Italy, I had applied for a Fulbright grant for sabbatical work on one of the most active plate boundaries on Earth, the Alpine Fault on the South Island of New Zealand. In addition to its tectonic renown, the west coast of the South Island is also one of the rainiest places on the planet. As a result, rocks recording the past seismic history of the Alpine Fault—including very young pseudotachylytes—have been exposed at the surface by ferocious erosion. The allure of those ravaged rocks carried the kids and me halfway around the world.

All of New Zealand, or, to use its Maori name, Aotearoa, is a geologic wonderland, with active volcanoes on the North Island and a still-rising mountain range on the South Island. Deep below the North Island, the Pacific oceanic plate is subducting westward beneath the Australian plate. The volcanoes there constitute a classic island arc, where water from the subducting slab reduces the melting temperature of the mantle rock above it, yielding gassy, viscous, and often explosive magmas. But this is the simplest part of New Zealand's tectonic story.

Just south of the South Island, there is a second subduction zone where the two tectonic plates have the converse relationship: the Australian plate is sliding eastward beneath the Pacific Plate. In between these two oppositely oriented subduction zones is the South Island, with the Alpine Fault as a kind of mediator. The long-term rate of subduction is slightly faster in the north than the south, however, and the Alpine Fault must accommodate not only both the inexorable convergence caused by the two subduction zones but also this north-south disparity in plate motions. As a result, the sense of slip on the Alpine fault is oblique: a hybrid of head-on collision and "strike-slip" displacement, like that of the San Andreas. The Southern Alps, whose craggy peaks stood in for Middle Earth in the *Lord of the Rings* films, are testament to this sideways squeezing.

Soon after arriving in New Zealand, I found I had to re-calibrate my sense for the rates of geologic processes. In Wisconsin, we are accustomed to well-preserved landforms from the Pleistocene Ice Age—glacial hills like sinuous eskers and teardrop-shaped drumlins are virtually unchanged from the time when the ice sheets melted back 15,000 years ago. But on the South Island of New Zealand, where internal tectonic forces are raising mountains by almost a half inch per year—and external processes of erosion are tearing it down at close to the same rate—the landscape is raw and young, subject to constant erasure. Few Pleistocene landforms survive, and rocks long buried in the crust are rapidly being exhumed. The energy in the New Zealand landscape is everywhere so evident—the swift streams, the terrifying landslides, and the pulverized fault rocks.

This sense of accelerated reality, of geologic events transpiring more rapidly than I had believed possible, paralleled a feeling that the pace of our personal lives had quickened, and that the landscape of our family was suddenly shifting. My oldest son would soon graduate from high school, and our tight little mother-bear-and-cubs world, the framework of our existence for so many years, would begin to vanish. Our time in New Zealand was our last great adventure together under that old mode.

Living in New Zealand felt to me like revisiting the best

aspects of the 1970s in the United States—a more liberal, creative, and optimistic time when there was still a shared belief in the value of public education and infrastructure. Dunedin had a thriving downtown district and an airy library that served as a lively hub of civic life. The public schools my kids attended were small and the atmosphere was relaxed; even at the high school, classes began at the civilized hour of 9:00 a.m. There was an ambient feeling of collectivity that reminded me of church suppers and county fairs in my childhood. A tiny observatory just down the steep hill from where we lived held open house nights once a month, with amateur astronomers describing special features of the Southern Hemisphere sky for attendees, who waited amiably in line to peer through the old telescope.

In preparation for a celebration of the winter solstice in June, scores of volunteers held workshops where children and adults could create huge lanterns from tissue paper and reeds, with wires inside to hold an actual candle. Beginners like us created simple pyramidal lanterns, while more experienced participants constructed giant flowers and fanciful beasts. After sunset on the night of the solstice, streetlights in downtown Dunedin were turned off, participants lit the candles in their lanterns, and thousands of people of all ages processed together in a glowing train on the darkest night of the year. The mood was simultaneously convivial and serious. I had been looking forward to the event but had not expected to find it such an emotionally powerful, raw human experience. Merging into a stream of people with flickering lamps in the vast dark night stirred some deep, atavistic pagan memory in me. It was extraordinary to experience the winter solstice in its own right, disentangled from the tentacles of Christmas.

Before we arrived in New Zealand, I had not fully appreciated the ruggedness of such geologically young terrain. The
distance from the University of Otago in Dunedin on the
east coast to exposures of the Alpine Fault on the west coast
was only about 130 miles. I imagined that once the boys were
established in their routines at school, I could zip over there,
make observations, collect some rocks, and get back again in
quick overnight trips—maybe even in a single day, if I got an
early enough start. In reality, the drive took about six hours in
good weather, and significantly longer if one encountered fog
or snow in the mountains. I was embarrassed to have underestimated the authority of the landscape.

Colleagues from the university did show me the wonders
of the Alpine Fault and also educated me in the peculiar sar-
torial style of New Zealand geologists: polypropylene long
underwear worn under shorts—the best strategy for staying
reasonably warm after fording waist-deep streams. They also
helpfully suggested a different study site in the central part
of the South Island, closer to Dunedin, where some unusual
pseudotachylytes—the frictional melt rocks formed in ancient
earthquakes—were exposed along an exhumed Cretaceous-
age fault zone, active around 100 million years ago.

Given that tens of thousands of earthquakes happen
around the world every year, including a dozen or so of magnitude 7 or greater, it would be reasonable to assume that
pseudotachylyte should be ubiquitous on ancient fault zones.
Besides, 90 percent of the stored-up energy released when
a fault slips in an earthquake is dissipated as frictional heat.
Most of the remaining energy is consumed in the fracturing
and grinding of rock along a fault as the rupture propagates,
and only about 1 percent becomes the seismic vibrations that

Black pseudotachylyte veins formed by frictional
melting during an ancient earthquake.

we experience as earthquakes. It's terrifying to imagine what
earthquakes would be like if friction didn't consume so much
energy (a thought that actually occurred to me in the seconds
after being awakened by P-waves in the early morning darkness in Italy).

Yet glassy pseudotachylyte, produced by frictional heating,
is comparatively rare on ancient fault zones—and this must
be telling us something important about the physics of fault
slip, how rock masses scrape past each other. The unexpected
scarcity of pseudotachylyte probably reflects several factors.
First, only seismic slip rates—on the order of several feet per
second—can generate enough frictional heat to melt rocks,
and slip on faults is not always seismic. The central section of
California's San Andreas Fault, for example, accommodates
differential motion between the Pacific and North American
plates through continuous "creep," rather than sticking and

then lurching in earthquakes. A second explanation for the dearth of pseudotachylyte along ancient fault zones is what geologists call "preservation potential": on faults that have experienced many earthquakes, early formed pseudotachylyte may simply have been ground up and destroyed by later slip, seismic or otherwise.

Another reason that pseudotachylyte is rare—at least according to a long-prevailing view among my fellow structural geologists—is that if there is water along a fault zone, any frictional heat generated in the first stages of fault slip will be consumed in raising the temperature of that water, which will then expand and separate rock surfaces from each other. This process of "thermal pressurization" reduces the area of frictional contact along a fault, which suppresses further heating, which precludes melting and the formation of pseudotachylyte. On a planet as watery as Earth, only previously unfractured crystalline rocks like granites or gneisses—or our Norwegian granulites—are likely to be dry enough, according to this logic, for frictional melting to occur.

But the rocks on the Cretaceous fault zone near Dunedin challenged that hypothesis. These rocks were turbidites—my dear old friends from summers in Ellesmere—whose entire life history involved water. They began as sediments deposited on the deep ocean off the coast of Australia and were later caught up in a Mesozoic tectonic collision that formed "Zealandia," a mostly submerged landmass (which New Zealand geologists argue should be considered a distinct, eighth continent). The hydrous metamorphic minerals in the rocks, including abundant micas, indicate that these sediments were awash in water in the middle crust at the time of deformation. In addition, there are veins cutting through the rocks, indicating that fluids carrying dissolved elements were coursing through fractures. Yet several fault surfaces in these watery rocks had also generated beautiful pseudotachylytes. How could this be?

I had collected a few samples from the fault zone on a preliminary visit but found their dialect cryptic. I needed to spend time with the rocks in situ to understand what they were saying, and a school holiday for the kids provided an opportunity. By that point my sons were well-seasoned field assistants. They had seen more pseudotachylytes than most geologists do in a lifetime, but it was unfair to ask them to spend their vacation in indentured servitude. So we rented mountain bikes, and they explored nearby trails while I tried to decipher the messages encoded in the ancient fault zone.

It often takes time for one's eye to become attuned to the palette of rocks. Subtle variations in color and texture are the key to their biographies. I spent most of an unproductive day taking measurements and notes without much feeling for how any of these observations fit together. I was feeling bad about neglecting the kids with little to show for my time. Then, late in the afternoon, when the sun illuminated the rocks from a different angle, I noticed that adjacent to the very black bands of pseudotachylyte glass—representing the primary fault surfaces on which melting had occurred—there were dark gray masses of ground-up rock that had apparently been intruded into the rocks outside the fault zone. Some of these masses contained fragments of pseudotachylyte, while others were cut by pseudotachylyte. This meant that the ground-up rock masses were both younger than *and* older than the pseudotachylyte—that is, the two had formed in alternation and represented a record of at least one full earthquake cycle.

Back at the university, I made thin sections of the rocks, and these confirmed the impression I had had in the field: fine-grained rock material produced by grinding on the fault zone had indeed made deep inroads into the enveloping rocks along a network of fractures—a telltale sign of thermal pressurization. During an ancient earthquake, superheated, overpressured fluids on the fault zone had shot into the surrounding rocks, carrying ground-up material from the fault zone with them, including some pieces of pseudotachylyte that must have formed in a previous earthquake. Once the fluids escaped in this way from the fault zone, the contact between the rock surfaces would have reestablished, generating a second generation of pseudotachylyte that cut across the ground-up rocks.

The rocks were telling us that, contrary to the prevailing paradigm, the presence of water in a fault zone does not always preclude frictional melting. As long as a fault zone remains permeable enough for heated fluids to escape, thermal pressurization and rock melting can *both* occur—even in a single earthquake event. The frictional properties of a fault can vary in a complex way even in the short duration of an earthquake.

This epiphany contributed to an emerging consensus that the events we call "earthquakes" are actually very heterogeneous, representing a wide range of different physical phenomena. Given the almost infinite number of possible combinations of rock types, fault geometries, hydrologic characteristics, and tectonic histories, no two earthquakes—like snowflakes—could ever be identical. We can abstract general principles about how and why they occur, but we will also miss something essential about them if we are too eager to

strip them of idiosyncrasy and turn them into abstract, idealized models. In the case of understanding earthquakes, a
preference for tidy idealization can even have deadly consequences.

While I was puzzling over ancient earthquakes in New
Zealand, geophysicists were documenting a previously unrecognized category of seismic events at modern subduction
zones. A new generation of broadband seismometers, capable
of "hearing" a wide spectrum of seismic tones, had picked up
unusually low-frequency waves emanating from subduction
zones, well below the depths where "normal" earthquakes
occur. These reverberations, called "seismic tremor," were
too gentle to cause any surface damage, but they could last
for weeks before going silent again for months. Meanwhile,
satellite-based monitoring of ground deformation revealed
that the seismic tremor episodes coincided with periods of
detectible slip along the subduction zone. These "episodic
tremor and slip" events were the first discovered examples of
what are now called "slow earthquakes"—at an inch or two a
month, livelier than ambient tectonic motions, but more lan-
guorous than conventional earthquakes, with slip rates of feet
per second.

The physical phenomena underlying these events—and
the rock record that they may leave behind—continue to
be the focus of much scientific debate, bringing the previously distinct fields of structural geology and earthquake seismology together in invigorating ways.[1] It's likely that we
structural geologists have been looking at evidence of slow
earthquakes for decades without knowing it. In retrospect,
it seems so obvious that there would be a whole spectrum
of fault behaviors between slow, continuous creeping and

sudden, violent lurching. Although our failure to recognize the phenomenon of slow earthquakes was due in part to a lack of instruments to detect them, it was also caused by a limited concept of what an "earthquake" was.

It occurs to me that our family, like most others, had experienced both fast and slow "earthquakes," and that the slow ones—long-term illness, unresolved grief, chronic stress—were in many ways more devastating for being stealthy, hard to name, and indefinite in duration. The same may be true at the collective level; we are quick to recognize need in times of sudden calamity, but we discount the power of slow cataclysms like grinding poverty, creeping corporatization, and eroding optimism to distort the character of an entire society over time.

Our inaction on climate change is another manifestation of the same disregard for the power of the incremental. Some evolutionary psychologists suggest that we humans are so hardwired to react to immediate threats that we simply can't make rational decisions over longer timescales. But I suspect this time-blindness is largely cultural, and most acute in the West.

Certain Western habits of mind have similarly shaped what is "seen" and what is ignored in the sciences. The late discovery of the full range of fault behavior is an illustration of how an understanding of Earth has sometimes been hampered by the intellectual preferences of Western science, which from the time of the Enlightenment onward has been dominated by the culture of physics. Among these preferences is the high value placed on "universality": veneration of simple, general laws that have the greatest possible explanatory power. In my undergraduate physics courses, we were taught to believe that physical laws, with their tidy mathematical formulations, were in some sense more real than any actual manifestations of them. Unruly entities like rocks, oceans, lifeforms, ecosystems—or a planet with all of these messy things—resisted such analysis and were by definition unworthy of consideration, mere "epiphenomena" whose study was left to lesser sciences.

Physicists often speak mystically of the "elegance" and "beauty" of natural laws and their mathematical descriptions, but to me this has always seemed to reflect a narrow aesthetic sensibility. An underlying assumption in physics is that simpler is always better, that uncluttered models free of bothersome things like friction—or history—are necessarily closer to scientific truth. While this is a powerful approach that has led to great technological advances, it is not the optimal way to study complex natural systems, which are often very cluttered—but nonetheless beautiful—and cannot be extracted from their moment in time without losing some inherent part of their character.

Stanford philosopher of science Helen Longino has proposed an expanded set of scientific values or "virtues" that enlarge the criteria for well-formed hypotheses.[2] Among these is the acceptance of "*ontological heterogeneity*"—embracing differences and idiosyncrasies as real and worthy of study, rather than mere deviations from a "standard." In Longino's words, "Ontological heterogeneity permits equal standing for different types, and mandates investigation of the details of such difference. Difference is resource, not failure." As Darwin recognized, tiny differences are the key to evolution. Without deviation and variation, there is no story. Earth is "governed" by the eternal laws of physics, but Earth's unique character

240 TURNING TO STONE

emerges from the profusion of things that it has invented by riffing creatively on those laws. Earthquakes—fast, slow, and in between—are among the innumerable examples of the planet's compositions.

The field course I was co-teaching in Italy was all about the growth of the Apennines, but the experience of the earthquake was a bit like studying the sun by looking directly into it. Even my unflappable Italian colleague, S, the resident geologist, was rattled. No one was able to sleep after the first tremor in the early morning, and the near-continuous aftershocks made it impossible to contemplate a normal day in the field. At dawn, I realized that in a few hours parents of the students would be waking up in the United States to news of the earthquake. I sent a quick, lighthearted email message to the American coordinator of the program, asking him to tell families that we were all unharmed and that the landscape had simply given us some real "experiential learning"—betraying none of the panic I had felt for my own children. The website of Italy's National Institute for Geophysics and Volcanology, INGV, reported that the epicenter of the mainshock was in a rugged part of the Sibillini Mountains in Abruzzo, about thirty miles to the south—so we decided to head in the opposite direction for the day, north to Urbino, where immersion in Renaissance art and architecture might bring reassuring calm.

On the way back, we stopped at a local vineyard to refill the field school's stock of verdicchio, a light white (or literally, green) wine that is a signature product of the Marche region. As we unloaded our large glass carboys to be filled directly from huge aluminum tanks, I noticed the family name on the wall: Beltrami. This was also the name of the county in northern Minnesota where my mom had taught high school on the edge of an Ojibwe reservation. I asked the owner of the vineyard in my bad Italian if any of her ancestors had any connection to Minnesota. She replied that yes, in fact a multi-great-uncle, Giacomo Beltrami, had been an explorer who claimed to have found the source of the Mississippi River in 1823. That would have been nine years before its "official" discovery by Henry Schoolcraft, the man who had also set off the copper rush in the Upper Peninsula of Michigan with his report to the War Department on the Ontonagon Boulder.

Signorina Beltrami went on to say that her ancestor Giacomo had had some help from Indigenous guides (who, I added silently, no doubt already knew where the river began). Then she mentioned that the family had many Ojibwe artifacts, including flint arrowheads, in a private museum that was unfortunately closed for renovation. This unexpected small-world historical connection stunned me. It was quite possible that my mother had had descendants of those guides in her classes, and now I had, improbably, discovered the source of Beltrami. The serendipity—and the wine—helped us stop thinking about the earthquake, *il terremoto*.

The next day, we were ready to continue with the geological curriculum, and I suggested to S that we replace one of the planned modules with a new one focusing on the earthquake, its tectonic context, and the importance of clear communication with the public about seismic risk. Just seven years earlier, a similar-size earthquake had occurred in the same region, leading to the deaths of more than three hundred people in the town of L'Aquila. Geophysicists around the world were stunned when six seismologists who had issued a report the year before on seismic hazards in the area were charged—and then convicted—of manslaughter for misleading the public about the likelihood of an earthquake. The case sent a chill through the geoscience community because earthquakes cannot be predicted in a deterministic way; geologists can only make probabilistic statements about potential risks to help inform building codes and infrastructure development. The L'Aquila case suggested that even recommendations based on the best available scientific information could be used as a basis for prosecution in court. Fortunately, two years after their conviction, an appeals court did clear the Italian seismologists of the charges.[3]

But S, who had taught the field course for many years, preferred to stick with the syllabus as planned. And he had a different narrative to share—about how Earth is dictated by external forces. The next project focused on "cyclostratigraphy": looking for the signature of astronomical cycles in sedimentary rock sequences. There is compelling evidence from ice cores that orbital variations collectively known as "Milankovitch cycles" (for a Serbian mathematician who worked out their complexities in the 1920s) play a role in long-term climate change. In particular, they seem to have modulated the glacial-interglacial oscillations during the Pleistocene. The Milankovitch cycles include variations in the ellipticity of Earth's orbit around the sun ("eccentricity"), the tilt of the Earth's rotation axis ("obliquity"), and which hemisphere leans toward the sun at different points in the orbit ("precession"). Each of these, over periods ranging from 19,000 to 400,000 years, affects the way in which solar radiation falls on Earth.

These orbital variations arise from complex gravitational interactions among Earth, the moon, sun, and other planets, and have presumably been going on throughout all of Earth's history. But it seems that there have only been certain intervals in the geologic past, such as the Pleistocene, when the Earth's climate system—and especially the carbon cycle—have been particularly sensitive to Milankovitch cycles. Stratigraphers have begun to look for evidence of these cycles in rocks from earlier periods in Earth's history, and the uninterrupted sequence of Mesozoic limestones in the Italian Apennines—the "Umbria-Marche succession"—is an ideal archive to study.[4]

Although these limestone beds have been wrinkled and imbricated by tectonic forces for more than 20 million years, we can reconstruct their original stratigraphic sequence. Students in our course were just learning the names and personalities of the various strata. At the base of the stack is the Jurassic "Calcare Massiccio," or massive carbonate, that creates the rugged spine of the Apennines. Built up by tiny calcite-secreting organisms on an ancient continental shelf, the Calcare Massiccio now houses the Gran Sasso ("Big Rock") Laboratory, the largest underground research center in the world, where physicists listen for elusive subatomic particles called neutrinos. I savor the fact that the particular, idiosyncratic, geological history of this rock—a "mere epiphenomenon"—makes it possible for the facility to seek timeless truths about matter.

Next are Cretaceous beds: the Maiolica Formation, an ivory-colored limestone named for its resemblance to Renaissance ceramic tiles. Above it is the Scaglia Bianca, a striking zebra-stripe unit in which white limestone alternates with thin layers of gray and black chert, or flint, the remains of tiny silica-secreting organisms called radiolaria. We're not the first to take note of those chert layers; ancient people knapped knife blades and scrapers from it. We've found some of these artifacts in recently plowed fields near our station. Like glass, chert is noncrystalline and can be honed to a sharp edge, making it a valuable and widely traded commodity in ancient cultures all over the world. Farther north in the Tyrolean Alps, Ötzi the Ice Man, the Bronze Age time traveler whose body was discovered in a melting glacier in 1991, was carrying arrowheads made of chert from Cretaceous limestones.[5] After the cherty Scaglia Bianca comes the Scaglia Rossa, a pink limestone that spans the end of the Cretaceous period—the last days of the dinosaurs—and the beginning of a new world in the Cenozoic. The delicate salmon color of the Scaglia Rossa defines the palette of Italian villas. Massiccio, Majiolica, Bianca, Rossa—miles of lime.

As a structural geologist, I'm new to cyclostratigraphy and learning along with the students. My colleague S divides the class into several groups and sends each to a different outcrop of the stripy Scaglia Bianca for the day. I drive with a group of four students to a roadcut on a steep mountain road with no shoulder and act as combination stenographer and traffic monitor as they measure the thicknesses of the repeating limestone and chert layers in the rock. This is a tedious process, and in the midday sun, the white rock is almost blindingly bright. The students call out measurements and I record them. "One centimeter limestone; one centimeter chert; three centimeters limestone; one centimeter chert." Over the next two hours we relive about a million years in the late Cretaceous.

The Cretaceous Earth was a greenhouse world; there were no ice caps, and both sea level and marine biological productivity were exceptionally high. Calcifying organisms flourished in the warm, expansive continental shelves, leaving behind great thicknesses of limestone; indeed, "Cretaceous" comes from the Latin word *creta*, meaning "chalk" or fine-grained limestone made of tiny shells. But Cretaceous limestones around the world, including the Scaglia Bianca, are punctuated by dark, sulfur-rich layers of chert, indicating that things episodically went amiss in those warm, prosperous seas. The black cherts record anoxic events: times when vast tracts of the ocean were oxygen-starved "dead zones," like the notorious one in the modern Gulf of Mexico, the result of excess nitrogen and phosphorous runoff from agriculture. The cause of the ancient low-oxygen episodes is not fully understood. Given the growing areas of anoxia in the Anthropocene oceans, and the threat these pose to global marine ecosystems, studying their Cretaceous analogs is more than just an academic exercise.

The biggest, baddest anoxic event of the Cretaceous is recorded by a sinister, sulfurous, organic-rich layer at the top of the Scaglia Bianca: the "Bonarelli level," recording a global anoxic event in the oceans about 94 million years ago. It reminds me, irrationally, of the first fish I ever caught in a northern Minnesota lake, a black bullhead—whiskered, rubbery, Mephistophelian—a species that lurks in low-oxygen waters. The narrow bands of black chert we are measuring in the Scaglia Bianca are mini-Bonarellis, thought to represent less severe swings in ocean chemistry, when lower oxygen conditions favored blooms of silica-secreting zooplankton (radiolarians) rather than calcite-producers (foraminifera). At our sun-baked outcrop, it's hard to tell at a glance if the variations are cyclic or random. To me, the alternating bands of limestones and chert in the Scaglia Bianca convey a sense of impending doom—a marine ecosystem that keeps veering dangerously toward collapse, then rights itself again, until finally something tips it over the edge and—Bonarelli. It makes me feel a little dizzy. I can empathize with the Cretaceous ocean; I know that feeling when things are about to spin out of control. Or maybe I'm just still wobbly from the earthquake.

At last we reach the top of the outcrop, pack up our gear, and wind our way back down to the field school to analyze the data. In order to recognize the signatures of Milankovitch cycles, which have well-known periodicities, we need to know how much time a given thickness of sediment represents. The known ages of the Bonarelli and other marker beds in the Scaglia Bianca place broad constraints on sedimentation rates, but our estimate of one centimeter per thousand years is rough. Students enter their hard-won measurements into software that can pick out sine-wave signals from noisy data. Our group's data appears to yield a 95,000-year signal, one of the characteristic frequencies of eccentricity, though it is predicted to have a fairly weak effect on how solar energy reaches Earth. To S, this is clearly the "right" answer—evidence that the chemistry and biology of the Cretaceous ocean was governed by astronomical factors.

That seems to be the lesson for the day: Earth is just a helpless marionette bobbling in space at the gravitational whim of other objects. I find the thought unspeakably depressing. It's late, we're all exhausted, and I don't want to seem to challenge S in front of the class, so I keep my questions to myself. If that really is an orbital signal, it was processed in some complex way by the Earth system—What exactly happened in the oceans to create the pattern of sediments that we see? Why didn't it respond to some of the other pulses of the

Milankovitch cycles? It seemed to me that Earth wasn't just dancing to an externally imposed beat; it picked up a rhythm it fancied and improvised on it.

I understand why S likes to emphasize the image of Earth as a rock pirouetting through space. He had been among the graduate students working with Berkeley geologist Walter Alvarez at the heady time when Alvarez first hypothesized that a meteorite had struck Earth at the end of the Cretaceous period, causing the extinction of the dinosaurs. This was a radical break with Lyellian "gradualism"—the geological aversion to catastrophic explanations—and represents one of the most important paradigm shifts in the history of the geosciences. As it happened, Alvarez and his group discovered the initial evidence for the impact event not far from our field station in the central Italian Apennines, near the medieval town of Gubbio. In fact, we had made a pilgrimage there just a few days before the earthquake.

The limestones of the Apennines were deposited at the time of the dinosaurs, but they contain no dinosaur fossils, since they are marine rocks, deposited on the continental shelf of ancient North Africa. They are, however, teeming with the tiny shells of single-celled organisms, especially foraminifera, or forams, which since late Paleozoic time have been among the planet's most important carbonate-secreting organisms. In the Apennines, the time of the dinosaur extinction is recorded by the unit that lies above the Bonarelli level, the pink Scaglia Rossa, and its forams are the ones who attest to the horrors of that event.

Seen under the microscope, the lower half of the Scaglia Rossa is like a densely populated, demographically diverse city, with forams of many shapes—tubular, coiled, lenticular, spiky, and globular, some resembling tiny raspberries. Then a distinctive layer of dark clay, less than an inch thick, marks a major change in the citizenry: Above this layer, forams are sparse, small, and far less varied in shape. Clearly, something terrible happened in the time represented by that thin layer of clay. This is the Cretaceous-Tertiary, or K-T boundary— the moment in geologic time when the dinosaurs and many other organisms, both on land and in the oceans, went extinct. (That moment is now more properly called the K-Pg boundary: In 1990, the International Stratigraphic Commission abolished the anachronistic "Tertiary" Period—a relic from an eighteenth-century version of the geologic timescale—and the Cenozoic now begins with the "Paleogene.")

In the late 1970s, when Walter Alvarez and his students were mapping the structural features in the Apennines, the distinctive K-Pg clay layer served as a useful marker within the thick Scaglia Rossa sequence, an aid in delineating the geometry of folds and faults. Alvarez became curious about the origin of the clay layer and how much time it might represent. His father, who happened to be the Nobel Prize–winning physicist Luis Alvarez, suggested measuring the concentration in the clay of a rare metal, iridium, that is delivered to earth mainly by dust-size micrometeorites. The background rate of this metallic "rain" had been recently determined from Antarctic ice cores, and the idea was that by determining the amount of iridium in the clay and dividing by the average rate of accumulation, they could determine how much time had elapsed during the deposition of the layer. What the Alvarezes found, famously, was a spike in iridium concentration for which there were only two possible explanations: either the clay layer represented millions of years during which lime-

stone precipitation mysteriously ceased and only microme-
teoritic dust accumulated—or a very large meteorite (in the
range of six miles in diameter) had delivered a lot of iridium
all at once.[6]

For ten years, geologists scoured the globe for a crater of
the right age and size, following the bread crumbs trail of
tektites—glassy blobs of molten rock from the impact site—
that were found to be present wherever the K-Pg boundary
layer was preserved. Tektite descriptions from sites around the
world were compiled, and it soon became clear that the number and size of tektites increased toward the Gulf of Mexico.[7]
Finally, in 1990, the mostly submerged crater at Chicxulub, on
the north coast of the Yucatan, was identified.

For many in the paleontological community, this deus ex
machina explanation for the dinosaur extinction event was
deeply unpalatable, a sci-fi scenario, a violation of Lyellian
decorum. Luis Alvarez had been involved with the development of the atomic bomb, and the parallel with meteorite
impact seemed distasteful. It didn't help when he told a *New
York Times* reporter (parroting Ernest Rutherford): "I don't
like to say bad things about paleontologists, but they're really
not very good scientists. They're more like stamp collectors."[8]

Today, more than forty years later, almost all geologists
agree that a giant meteorite did strike Earth at the end of
the Cretaceous, and no doubt wreaked havoc with global
ecosystems. Yet there continues to be opposition to the idea
that the impact was the sole or even primary cause of the
extinction. The most vocal skeptic is Princeton University's
Gerta Keller, who has argued for years that the stratigraphic
record of the K-Pg boundary in the Caribbean region, close
to the site of the impact, suggests that the extinction was a

prolonged ordeal that may have been set in motion by one of the great flood basalt events in Earth's history, the eruption of the Deccan Traps in India.[9] Gas emissions associated with flood basalts are known to wreak havoc with global ecosystems; the Permian extinction, the worst in Earth's history— and the event that defines the end of the Paleozoic Era—is attributed to the knock-on effects of an earlier generation of flood basalts, the Siberian Traps.

The extinction could well have been caused by the combination of turbocharged volcanism and a rogue space rock, but the advocates for the two scenarios refuse to cede any ground to each other. The ongoing debate between the Alvarez camp and Keller and her cohort has reached unprecedented levels of vitriol, with name-calling ("gadfly," "crybaby," "bully," "bitch") and charges of scientific malfeasance on both sides.[10] As an outside observer, I wonder what it is all about. For some reason, dinosaurs seem to inflame the brains of many people, but given the bitterness of the feud, there must be other, higher stakes involved. There is no doubt that the Alvarez hypothesis was an astonishing example of inductive inference. It forever changed the rules for the kinds of natural phenomena that geologists could invoke to explain what we see in the rock record. Advocates for the meteorite impact theory can rightly point to that revolutionary contribution to the geosciences. But why the bilious attacks on Keller and others who suggest there could also have been internal factors that contributed to the extinction? And why do Keller and company reciprocate with equal venom?

It's hard not to sense some role of gender in this debate. The impact hypothesis merges Occam's razor and Thor's hammer: simple, absolute, unsubtle. Bam—the dinosaurs died!

The scenario has the punch of a comic book and the lure of a slasher movie; we can imagine the event playing out in real time and feel the pathos of a world suddenly destroyed. Keller's argument, in contrast, is far more nuanced; her description of the last days of the Cretaceous reads more like a Shakespearean tragedy: The dinosaurs struggle through a series of misfortunes and travails that in combination leave a devastated world. Ultimately, I think, the battle seems to be about the essential character of Earth and how we tell the narrative of the planet's history: To what extent has Earth had agency over its evolution versus simply reacting to externally imposed conditions?

In studying the Earth through time, it's so tempting to look for a coherent narrative arc, a tidy throughline, a moral of the story. But there are so many characters and intersecting plotlines that it's hard even to identify the appropriate genre: Gothic horror? Edenic idyll? Comedy of errors? A march of progress—or purposeless meandering? It's all of these and far more, a Borgesian library of stories in stories in stories. Or perhaps looking for narrative blinds us to the real nature of Earth. Maybe Earth's essence is simply its exuberant improvisation and continual reinvention, a never-ending jam session of rock, water, and life.

Although I'm an agnostic in the K-Pg debate, visiting the clay layer at Gubbio did feel like encountering a celebrity. The meteorite-meets-dinosaur story long ago entered the realm of pop culture. As I took a tiny sample of the famous clay, I thought of how the whole iridium/tektites/Chicxulub saga had put two retired high school biology teachers from Thunder Bay, Ontario, on the trail of one of the most astonishing recent discoveries in North American geology.

The teachers, Bill Addison and Greg Brumpton, knew that the ancient sedimentary rocks around Thunder Bay on the western shore of Lake Superior were about the same age, at around 1.9 billion years, as the giant impact crater at Sudbury, Ontario, more than four hundred miles to the southeast. The Sudbury crater is the second largest known crater on Earth, just a bit smaller than the Vredefort structure in South Africa, which was formed at close to the same time. The Sudbury crater was squashed into an elliptical shape by an ancient mountain-building event, but was originally 160 miles in diameter—significantly larger than the Chicxulub crater, which is 110 miles across and ranks third in size. The Sudbury crater is host to one of the world's largest nickel deposits—a gift of the meteorite—and is also a world-class monument to the devastation of acid rain, the legacy of more than a century of smelting sulfidic nickel ores. Physicists have repurposed the deepest underground mine at Sudbury as a neutrino laboratory similar to the one in the Calcare Massiccio at Gran Sasso.

Addison and Brumpton posed a question that no professional geologists had thought to ask: Was ejecta material from the Sudbury impact preserved in rocks far from the crater? This was the converse of the Alvarez hypothesis, in which the ejecta layer was discovered before the crater. The two teachers contacted geologists at the local university, who began scouring outcrops of the Gunflint iron formation around Thunder Bay for tektites and other telltale signs of impact. The Gunflint, like other banded iron formations in the Lake Superior region, record the early Proterozoic "Great Oxygenation Event," a great inflection point in the history of the planet—the transition of Earth's atmosphere from mainly carbon dioxide and water vapor to having some free oxygen (O_2), thanks to the tireless efforts of photosynthesizing microorganisms.

The early Proterozoic iron formations—from which virtually all the steel in the world has been forged—have sedimentary textures very much like limestones today, indicating that they were deposited in similar shallow continental shelf environments. In today's oceans, iron is vanishingly scarce—indeed a limiting nutrient for marine ecosystems—but the immense thickness of the Gunflint and other iron formations indicates that at one time, iron was a major constituent of seawater. This could only have been possible in oceans that lacked oxygen, because in the presence of oxygen, iron "rusts" out of the water as iron oxide minerals like hematite and magnetite. The iron formations record a transitory time in Earth's history when iron released from deep-sea volcanic vents could remain in solution until it encountered oxygen in shallow waters where photosynthesizing organisms thrived.

The old Gunflint is ribboned with layers of gray chert in a way that is reminiscent of the much younger Scaglia Bianca limestone. In both rocks, the cherts mark cyclical changes in ocean chemistry. In fact, the Gunflint is named for the chert that alternates with its iron-rich layers and was used in the days of the fur-trading voyageurs as the ignition mechanism in flintlock rifles. It could well have been the source of the chert arrowheads in the Beltrami collection in Italy. The Gunflint cherts are also famous in the annals of geology. In 1953, when the idea of Precambrian life was not widely accepted, paleontologists from Harvard and the University of Wisconsin found well-preserved fossils of single-celled prokaryotic organisms in specimens of the chert in northern Minnesota.[11] For decades (until the discovery of 3.4-billion-year-old bacterial fossils in Australia in the early 1990s), these were the world's oldest known undisputed microfossils.

So the venerable Gunflint had been well examined, but no one had ever thought to ask it about the Sudbury impact. The main challenge was determining where in the iron formation sequence to look. Fossils in the Gunflint are too sparse to provide a biological record of the event in the way that foraminifera mark the K-Pg boundary in the Scaglia Rossa. Still, the university geologists were intrigued by the teachers' suggestion, and they hoped that high-precision dating of zircon crystals in volcanic ash layers within the iron formation could make it possible to zero in on the level within the Gunflint that corresponded to the time of the impact at Sudbury.

A PhD student took on the zircon dating project and identified where in the Gunflint the ejecta layer should be. Geologists were stunned to find that it had been under their noses for years—including in outcrops in the center of the city of Thunder Bay.[12] It's hard to comprehend now how no one had noticed these before: They contain not only abundant tektites but also large chunks of iron formation lying in random orientations, apparently ripped up by a tsunami that the impact unleashed. Furthermore, the layer contains odd, concentrically layered pellets, interpreted as "hailstones" that formed around water droplets as they fell through an atmosphere thick with ash-like pulverized rock. Strangest of all are glassy, spongy masses that look like holey, vesicular basalt but appear to have been formed by rock that was so hot it was a vapor that *boiled as it cooled*.[13] There is no nuance or understatement here; this is extreme, in-your-face geology, as shocking as encountering a shootout in the pages of Jane Austen.

Geologists began to find the distinctive ejecta layer, with its tektites and pellets, in banded iron formations elsewhere in the Lake Superior region—a record of one particular very bad day 1.85 billion years ago. It has even been possible to reconstruct the events that occurred that day hour by hour. In the area that is now Thunder Bay, ground shaking that was equivalent to a magnitude 11 earthquake (larger than any fault can generate) would have been followed ten minutes later by a ground surge of ejecta traveling at more than five hundred miles per hour, then a supersonic air blast, and finally a titanic tsunami.[14]

None of this would have been discovered if a couple of high school teachers hadn't nudged some geologists in the right direction. Addison and Brumpton were made first authors on a series of papers related to the impact (including several cited above) and became familiar faces at the annual Institute on Lake Superior Geology meetings and field trips that my students and I attend each year. Searching for additional ejecta layer sites became an irresistible treasure hunt for geologists in our region. Colleagues I know from the USGS collected a boulder-size piece of the ejecta layer in the Upper Peninsula of Michigan to bring to the Smithsonian, and on the long drive to Washington, DC, they nearly had an accident. The huge rock was in the back of their SUV, and when the driver suddenly hit the brakes to avoid a deer, the boulder rolled forward and almost struck them. They later joked that they could have been the last victims of the Sudbury impact.

Sampling the clay layer on our pilgrimage to Gubbio, I think about the contrasting stories of how the K-Pg and Sudbury impact events were reconstructed. One started with an ejecta layer that pointed to an unknown crater; the other began with a crater that implied the existence of an unrecognized ejecta layer. One involved a Nobelist and his academically distinguished son, the other started with a hunch two school-teachers shared. The most striking difference, however, is the bonhomie associated with the discovery of the Sudbury layer and the bitterness and bellicosity connected to the K-Pg boundary. In my experience, rocks tend to instill humility in people who spend time with them; they remind us that we are only partly fluent interpreters attempting to translate complex texts. But sometimes human egos become entangled with particular translations, and any nuance in literary analysis is lost.

I think also of the ancient people who knew these rocks, on opposite sides of the ocean, as sources of chert. A few weeks earlier, we had made a trip to the Dolomites and visited Ötzi the Ice Man in the Archeological Museum in Bolzano. The exhibit of objects Ötzi had with him emphasized the cultural continuity between this Bronze Age man and modern Tyroleans. He used some of the same basic tools as people in the region did as recently as the early twentieth century and ate foods that are still gathered or cultivated today. The display celebrated the unbroken chain of more than 250 generations that connected this man with people now living in the area. I was struck by the contrast between that sense of connection with the past and the feeling of loss and shame, of complicity in apocalypse, that I have in viewing Native American artifacts like the ones kept in the Beltrami collection. From that vantage point in the Old World, the enormity of the devastation to indigenous North American cultures hit me anew, like a shock wave. Vast knowledge of rocks and plants and ecosystems, rich lexicons to describe them, and whole ways of living on Earth are recorded now only by mute flint spearpoints.

I turn back to the rocks. It's less distressing to contemplate geologic apocalypse than genocide. The Sudbury event was so violent that it must have had an effect on early Proterozoic life, but the fossil record is too scant to tell if there was a mass extinction. The biosphere was entirely microbial at the time, and perhaps more resilient than later ecosystems with elaborate food chains—no *T. rexes* with their precarious dependence on carnivory. All around the Lake Superior region, the Sudbury ejecta layer lies at the top of the iron formations and is overlain by dark black sulfurous shales, suggesting, at least, that the chemistry of the ancient oceans in which the iron formations accumulated was dramatically altered in the aftermath of the impact.

My students, my sons, and I have bushwhacked and braved wood ticks looking for exposures of the Sudbury ejecta layer in Wisconsin, to no avail. Although it does not crop out at the surface, we know exactly where it should occur at depth. The layer has been encountered in drill cores, but in a strange plot twist in my own life, I have not been allowed to see them. The samples were taken by an out-of-state mining company known mainly for mountaintop-removal projects in coal country. The company declared intentions to blast a huge open-pit iron mine at the highest part of the Gogebic Range, a wild and beautiful line of hills in northern Wisconsin. In this area, the iron formation is tilted at an angle of 60 degrees to the north, toward Lake Superior, and mining the low-grade ore would have involved blasting and stockpiling a huge volume of the overlying black shales, which are full of the sulfide mineral pyrite. The scale of the proposed mine and amount of waste rock that it would produce were so unprecedented in Wisconsin that the state legislature had to change mining regulations for the project even to be legal.[15]

The site was near the headwaters of a river that flows into Lake Superior and supports one of the last extensive wild rice stands in the region, the Kakagon Sloughs, on the Bad River Ojibwe reservation. Wild rice is particularly sensitive to pH, and any acid drainage from the sulfidic waste rock could devastate the sloughs. After so much time spent looking for the Sudbury ejecta layer, I knew the rocks at the proposed mine site intimately. And because I was an employee of a private college, not one of the public universities or state agencies whose funding was controlled by the legislature, I was in a unique position to speak out.

I prepared a report for the Great Lakes Indian Fish and Wildlife Commission outlining the potential environmental effects of the mine and testified at the state capitol against a bill that was written to exempt the project from existing statutes. This naturally made me a persona non grata in the eyes of the company, and there was no chance that I would ever be allowed to see the drill core with the ejecta layer. The company held the mining rights and paid for the drilling. By law, the rocks belonged to them. But I felt like a scholar who had waited for years to study a manuscript in a rare books library only to be barred at the doorstep from entering. The company didn't own the story of what happened 1.85 billion years ago. That belongs to all of us.

Before our semester-long field course in Italy is over, there are several more major earthquakes, *terremoti*, one of them even larger than the traumatizing event in August. This is a violation of the usual pattern of exponential decline in size and magnitude—aftershocks are expected to be less severe than the initial quake that they follow—and another reminder of the naivete of our simple-minded notions about how Earth works.

My mind is swirling with the experiences of the previous months. We've witnessed competing accounts of the character of the planet. Is it shaped more by violence—earthquakes and impacts, recorded by glassy pseudotachylyte and tektites—or by patience, embodied in limestone and iron formation and flint? Is Earth simply a hapless rock hurtling through space, prone to gravitational perturbations and collisions, or are its own tectonic habits and biogeochemical cycles enough to protect it from permanent damage? Are there elegant mathematical laws governing its behavior, or is it foolish to imagine that anything other than probability controls its destiny? Are the patterns and connections we discern in records of the past meaningful, or simply stories we tell in the hope that there is some order and logic in the world?

It occurs to me that the same questions apply to individuals and families, and the answer in every case is yes, both, all of the above. Our lives are simultaneously predictable and probabilistic; the stable ground beneath us sometimes shudders. Worlds that once seemed permanent vanish slowly—or suddenly—giving way to new ones. Mother Earth abides, but mortal mothers cannot shield their cubs forever. We can only gird them with life experiences and fortify them with love.

For now, the earth is still again, and all is well. My sons are fine; they're less shaken up than I am by the earthquake, and in a few days they'll leave for further adventures in Rome and Florence and Genoa. We are glass and flint. We are fragile, and we will endure.

10

QUARTZITE

Another April in Wisconsin, and the air is heavy with the smell of soil reawakening: organic, fecund, mysterious. The nerdy, overeager student part of my brain volunteers that it's the compound geosmin, literally "earth aroma," and specifically $C_{12}H_{22}O$, exuded by soil bacteria and fungi. But right now, I just want to close my eyes and let the scent carry me to Aprils past: It's the smell of playing marbles at recess in the schoolyard amid melting banks of snow; of waiting outside the church in our flowery Confirmation Day dresses, hoping we will remember our assigned sections of Luther's Catechism; of pushing my first son in a stroller on a trip to the library one Saturday in spring. I open my eyes again, feeling a little lightheaded, unmoored in time, and am glad to be on this hillside in the stabilizing presence of the old Baraboo Quartzite, a rose-colored rock that peeks through the blanket of sandstone in southern Wisconsin.

I'm on a field trip with the latest batch of majors in my structural geology class. When I began teaching, I was younger than some of my students. Now all of the students are younger than my children. I think of the many times I've been at this special spot, with my own professors and mentors, my sons, colleagues, and countless students on other trips that now blur into composite memories. The smell of the spring air has made me keenly aware of the passage of time, and I realize with surprise that I'm close to crying. I take a deep breath and get a grip on my emotions; it would be awkward to explain to the group that visiting this outcrop of pink quartzite is for me a sacred ritual. For most of them, this trip is an imposition, a disruption of their weekend plans.

They're a nice bunch of kids but unaccustomed to experiences not mediated by cell phones and social media. I want them to engage with the rock, in a raw, sensory way. I lie down on a bed of quartzite tilted about 30 degrees from horizontal to emphasize the fact that these rocks have seen some tectonic action. I'm sure this is deeply uncool. A couple of students crane their necks to signal interest, but they're all standing too far away to gather what the rocks have to say. I tell them to come close, put their noses right up to the rock face and tell me what they learn. They approach self-consciously, hands in pockets, embarrassed to appear too enthusiastic.

This particular exposure of Baraboo Quartzite is so famous to geologists that it has a proper name: Point of Rocks. For generations, geologists from around the midcontinent have made pilgrimages to study the rosy rocks of the Baraboo Hills—ancient sandstones recrystallized into durable quartzite, the eroded remnants of a mostly buried mountain belt of folded rocks. Some of the foundational concepts in structural geology—how small-scale features can be used to understand regional fold geometries—were worked out here by Charles Van Hise, a late-nineteenth-century geologist who later served as an influential president of the University of Wisconsin. This is also a good place to learn to read sedimentary

Point of Rocks, Baraboo Hills, Wisconsin

rocks; despite being deformed and recrystallized, the Baraboo Quartzite still remembers the tides and currents that deposited it more than 1.5 billion years ago.

After a few minutes in the company of the quartzite, my students begin to pick up its idiom. Two of them exclaim that they've found an amazing bedding surface—it's corrugated, ribbed. I let them imagine that they're the first to discover it. It's my favorite thing about this outcrop; the ancient ripple marks are perfectly shaped, undulating, almost sensual—sculpted by water that flowed a billion and a half springtimes ago. Michelangelo would have admired them. Soon everyone has come to marvel at the ripple marks, and like so many previous pilgrims to this spot, they can't help but stroke them in reverence. It makes me inordinately happy to see digital-age twentysomethings fall under the spell of these ancient sediments.

The rocks here in the Baraboo Range—a ring of oddly rugged hills in otherwise flat terrain—are old, but what is even more remarkable is that the landscape itself is old. In fact, it is one of the few places in the world where one can walk on a Precambrian land surface. The Baraboo Hills, made of the exceptionally resistant Baraboo Quartzite, represent "fossil" topography—part of a mountain belt that was buried for hundreds of millions of years under a succeeding genera-tion of sand. That blanket of sand was the Cambrian Sauk sequence, the sandstone that defined the world of my child-hood.

After a half billion years in the underworld, the ancient Baraboo mountains are now being "exhumed" by erosion; the sandstone cover is wearing away, and the quartzite hills bask once again in the open air. The contact, or boundary, between the quartzite and sandstone is the Great Unconformity—broadly the same erosional surface that separates the layered strata of the upper sections of the Grand Canyon from the famous Vishnu Schist at its base. The Vishnu Schist and Baraboo Quartzite, are in fact about the same age, cousins in time. In the Grand Canyon, the Great Unconformity is seen only in cross section in the steep walls, but at Baraboo, the entire modern land surface hews closely to the unconformity, revealing the ancient terrain in three dimensions.

I've always found it intriguing that the Baraboo Hills in-spired two central figures in the environmental movement in the United States: John Muir, who grew up on a farm nearby, and Aldo Leopold, who wrote *A Sand County Almanac* in a shack in the shadow of the Baraboo Range. I don't think either would have known the details of the geologic history, but perhaps they sensed the deep mystery in the primordial landscape. This parable of reincarnation and endurance is what I hope that my students will hear with their ears pressed close to the rocks. If the rocks spoke Italian, I'm sure they would quote the closing words of Dante's *Inferno*: *E quindi uscimmo a riveder le stelle*—"And so we came forth, and once again beheld the stars."

The students are a little overwhelmed; this is a complex story (even without invoking Dante), and only fragments of it can be read at any single spot. I can give them the narrative framework, but each must independently develop the capacity to map what they can see at a given outcrop—filtering out vegetation, human infrastructure, and other distractions—onto the appropriate coordinates in time and paleogeographic space. That is, they must train themselves to hold multiple versions of the world simultaneously in the mind's eye. I remind myself to be patient, that this is difficult, a metacognitive ability that is acquired only through practice. It's hard for the students to see the grand arc of this saga when they are still learning to distinguish sedimentary features in the rocks from structures that formed during the mountain-building process. They struggle to use their compasses to measure the orientations of bedding planes and faults. They can't keep the ages of the various events straight.

This is all perfectly normal. But today, I'm having a small existential crisis and would rather talk about the philosophical significance of these rocks. My Lutheranism is so long lapsed that it provides me no midlife solace. I think again of Confir-

mation class—con*for*mation, one of my friends called it—and
Luther's Catechism, written for young people preparing to
affirm their faith. After each of the Commandments and the
Apostles Creed, Luther asks: "What does this mean for us"?

If I never found Luther's answers very compelling, I now
see the wisdom of his repeated question. I want the class to
join me in an exegesis of the Baraboo rocks: What do they
mean for us? I'd like to discuss how our exploration of the
rejuvenated, once-and-future Baraboo Hills exemplifies Mircea Eliade's concept of the "eternal return"—the experience
of time in religious ritual, a counterpoint to the everyday perception of time, which he termed "profane duration." Eliade
wrote that in rituals, "Primordial mythical time is made present . . . indefinitely recoverable, indefinitely repeatable." Rituals reflect the deep human desire "to become contemporary
with the gods, to recover the strong, fresh, pure world that
existed *in illo tempore* [in those days]."[1]

As my students and I trace the Great Unconformity, internalizing the contours of the ancient Baraboo mountains
through our feet, it occurs to me that recovering the "world
that existed in those days" is exactly what we are doing. But
I won't tell them right now that understanding the "eternal
return" is the point of all the maps, cross sections, and field
notes they are to turn in after the trip. They need to discover
that for themselves.

After teaching for more than three decades, my priori-
ties for student achievements have changed; my younger self
might have said they've grown too lax. I've become more open
to tempering pure *logos*—analytical thought—with a bit of
mythos—the larger stories that knit the world together. I'm
less concerned than I once was about students acquiring a

wide vocabulary of minerals and rocks or mastering certain
field and lab techniques. Instead, my hope is that they develop a visceral sense for the way that the planet works—
for its sacred, indefinitely repeating rituals: the circadian
rhythms of the wind, the seasonal inhalation and exhalation
of the biosphere, the decades-long *hajj* of groundwater, the
millennium-scale overturning of the oceans, the global con-
tradance of biogeochemical cycles, the tectonic reincarnation
of rock through seafloor spreading and subduction, mountain
building, and erosion.

Although Earth, among the rocky planets, had certain advantages from the start—size, distance from the Sun—these
practices of endless re-creation are the true secret of its long-
term habitability. Ritual is what distinguishes Earth from its
lifeless siblings. I want students to perceive of themselves as
Earthlings, wholly dependent on the planet's sacraments—
and to realize that any *rational* society would align its practices with those of the system that sustains it.

If I do not always succeed, perhaps I can be forgiven. By
the time students come into my classroom, they have long
since absorbed prevailing societal conceptions about the relationship between humans and Earth. Anthropologist Clifford
Geertz famously defined culture as the constellation of stories
that groups of humans tell themselves about their place and
purpose in the world. In the United States, where capitalism
has become the de facto religion, the stories we tell about who
we are mostly exclude Nature. The natural world is simply a
passive backdrop against which the "real" stories unfold. So
when I try to convince students that the soils they've never
thought about, the natural landscapes that are obscured by
urban infrastructure, the rocks they assume to be deaf and

dumb, are actually in charge of everything, it's no wonder that they look away, unconvinced.

In the Western world, our shared rituals are no longer sacred but merely transactional—buying, selling, influencing, tweeting, endorsing, sponsoring, lobbying, and above all, *consuming*. At this point, it is hard to find our way back; our worship of novelty is in fundamental opposition to the "eternal return." Most of our technology is predicated on subverting the Earth's rituals rather than participating in them; we cannot become "contemporary with the gods" if we believe we've replaced them.[2] The cosmos is now the global economy, and the titans who rule it believe they are exempt from natural law. Some of them imagine that humans are destined to live on Mars—that we could transform a planet with no oceans or soils or tectonics into an Earth-like Eden in a matter of decades—or that we will, for some unexplained reason, be better off in the "metaverse." Even if we could homestead on a new planet, or wanted to live in a virtual world, we would still be us—the same flawed creatures expelled from the first Eden.

We understand far more about Earth than we did when I first became a geologist, yet we seem no closer to wisdom about ourselves. When I finished my PhD, my advisor gave me a biography of Charles Van Hise,[3] the structural geologist who mapped the Baraboo Hills and would have known this very outcrop. The volume sat on my bookshelf for thirty years before I finally read it and became a Van Hise groupie. Van Hise was a colleague of T. C. Chamberlin, the eponym of our Svalbard valley, who also served as University of Wisconsin president. Both geologists felt called to public service because they believed that science would lead society into a more enlightened age. Van Hise used his post as university president to advance strikingly progressive causes: not only public education but also conservation, women's suffrage, and limits on corporate power. He was a friend of Wisconsin's populist governor "Fighting Bob" LaFollette, and together they promoted the utopian "Wisconsin Idea": a vision for workers' rights, income redistribution, and access for all citizens to the resources of the university.[4] Today, that would be labeled a socialist agenda. Van Hise believed it was the path that a scientifically advanced society would logically follow.

I'd love to spend an afternoon with Van Hise on this outcrop and catch him up on what we know about the geology of the Baraboo Hills, rock deformation, and plate tectonics, but I'd be reluctant to tell him how far we are from his vision of the future. And I wouldn't share with him my conclusion, based on my own career in academe, that knowledge and rational thought alone cannot leverage social change—without an accompanying cultural revolution. Free-market economics has permeated too deeply into Western habits of mind, making behaviors that are considered pathological at the individual level acceptable at the scale of society as a whole. In fact, the so-called dark triad of personality traits linked with borderline psychopathy—self-centered grandiosity, calculated exploitation, and lack of emotion[5]—are, in the corporate world, just good business sense.[6]

I sometimes wonder whether, at this point in the history of human civilization, it would be possible to create a fresh new kind of secular spirituality free from both the narrow orthodoxies of traditional religions and the venal dogmas of capitalism. The primary texts would be the rock record and the book of nature; lullabies would reassure infants that they were in the care of a wise old planet; children would grow up knowing about planetary superheroes like carbon-gobbling dolomite and plate-moving eclogite; holidays would celebrate sandstone aquifers and stable granitic continents; the central principles would be wonder, gratitude, connectivity, collectivity. After a few generations of such reacculturation, an egalitarian ethos would emerge as we came to think of ourselves as Earthlings with deep bonds of kinship with one another, and all components of nature. Humans would aspire simply to blend in. Within this worldview, amassing disproportionate wealth, oppressing other humans, or degrading the environment would be seen as both unnatural and immoral. My utopian reveries are interrupted by a cynical voice in my head: *Yeah right, fat chance.*

Luckily, my ruminations on the state of civilization are interrupted at this point by the students, who are now clambering over the rocks together and have discovered another kind of distinctive feature on the surface of a bed in the quartzite. "Are these slickensides?" one asks, and my heart swells. Slickensides are linear features that record the direction of sliding on faults; their name is almost comically onomatopoeic. "Yes!" I say. "And they show us that the bedding planes in the quartzite served as slip surfaces—faults—at the time the rocks were folded." I demonstrate by bending my softcover field notebook and ask them to consider how it is possible to do that when the pages themselves have not been internally deformed. The students each perform the same simple experiment and see that the folding is accommodated by slip between the pages. In a sequence of layered rocks, bedding surfaces with low cohesion behave the same way when tectonic stresses cause the strata to buckle. This is how the strong, stiff quartzite was able to bend into the giant canoe-shaped syncline that defines the Baraboo Hills.

We climb to the top of Point of Rocks and see more evidence of this "flexural slip" process. Sandwiched between two thick beds of pink quartzite—erstwhile sandstone—is a dark, clayey layer, about three feet thick, that itself contains a thinner layer of quartzite. This was originally a sheet of mud within the otherwise sandy sediments, recording a period when the crashing waves that deposited these rocks on an ancient shore were subdued for a time. Tens of millions of years later, that brief becalming led to dramatic contrasts in mechanical behavior as the rocks buckled into great folds. The weak clay layer—like the air between the pages in my notebook—accommodated flexural slip between the two enclosing beds of unyielding quartzite. Unlike the air, which was unscathed in the process, the clay experienced intense and permanent deformation by shearing. This in turn caused the thin stringer of quartzite within the clay to become involuted in a dramatic, almost violent, way. A student who has hardly spoken in the weeks since the course began exclaims, "Whoa! Look at that!" This time I don't need to tell the group to come closer to the outcrop; they're already huddled together, trying to understand what these rocks are saying.

Imagining that one can converse with rocks is of course taboo in Western science, which values dispassionate investigation and preservation of a clear boundary between observer and observed. That much-vaunted objectivity could be compromised by developing any sort of relationship with one's subject. I suspect, however, that most geologists are secretly in love with rocks.

How interesting, given the epistemological differences between physics and religion, that it is likewise taboo in mono-
theistic traditions to speak of agency in the natural world;
such a notion undermines the authority of God. But according
to my youngest son, K—who, despite growing up in a thoroughly agnostic family, earned a divinity school degree—there
are hints of rock veneration in the Old Testament. When the
Israelites were close to entering the Promised Land after forty
years in the desert and nearly dying of thirst, Moses asked
God for help. God told him to gather the people before a
particular rock then: "Speak to that rock before their eyes and
it will pour out its water."[7] Moses led the desperate group to
the site and struck the rock with his staff. It yielded abundant
water, saving them and their cattle. But God then informed
Moses that he would be barred from the Promised Land because he failed to address the rock. This passage is usually
interpreted as a stern lesson about obeying God, but I choose
to think that God was also asking Moses to respect the rock
as a sentient being.

Puzzling over the wildly contorted structures at the top of
the outcrop, my students are beginning to perceive the rocks as
pliant, animate, responsive. The existence of an ancient mountain belt in southern Wisconsin is no longer an abstract notion;
they are deep inside it. They can grasp its scale and architecture. One student asks about the larger tectonic context of the
Baraboo mountain-building event. It's an excellent question.

I explain that despite the geologic fame of the Baraboo
Range, we really don't know much about why or even exactly
when these mountains formed. We know that the heart of
the Baraboo Range must have been to the south (in modern
coordinates), in what is northern Illinois, because quartzite

units similar to the Baraboo are found elsewhere in Wisconsin and they are less deformed toward the north. But the rest of the ancient mountain range still lies slumbering beneath thousands of feet of sandstone, shale, and dolostone, waiting to be revealed again someday by erosion.

The age of the deformation is also elusive, largely because of the nature of the quartzite itself. The sandstone from which the Baraboo quartzite formed is among the most chemically "mature" sedimentary rocks ever documented—at the time it was deposited, its constituents had been weathered and winnowed until nothing soluble or prone to breakdown remained. It consists almost entirely of quartz (SiO_2), with traces of iron that tint it pink, as well as some aluminum-rich clays, and that's all. The stripped-down simplicity of its composition is in fact the reason that these ancient hills have persisted across the eons—they are physically tough and chemically almost immortal.

This elemental austerity also makes the quartzite "timeless" in a literal geological sense—it didn't have enough ingredients to cook up metamorphic minerals that can be dated isotopically. Less-mature sedimentary rocks contain radioactive elements like potassium and rubidium, and when such rocks are involved in tectonic events, these elements may be incorporated into new minerals like micas that can be used to determine the time of heating and deformation. But the main effect of metamorphism on the super-winnowed Baraboo sands was simply to recrystallize the quartz into larger grains, fusing them together into a nearly indestructible, and undatable, mass. I rather admire the Baraboo's taciturnity, its resistance to our probing and analysis; the rocks seem to know when we need a good dose of humility.

The Baraboo Quartzite itself must be younger than 1.71 billion years old, based on the ages of the youngest grains of zircon that are sprinkled sparsely among all the quartz.[8] Until quite recently, geologists had no firm basis for knowing whether the Baraboo mountains formed tens or hundreds of millions of years after the rocks' deposition. The magnitude of the uncertainty was reduced dramatically by the discovery, in an abandoned mine shed, of old drill cores through a layer of slate that lies on top of the Baraboo Quartzite. The slate, a far less hardy rock than quartzite, isn't exposed at the surface, but it records information that the quartzite does not: Fine-grained micas in the slate yield metamorphic ages of 1.45 billion years—younger than many geologists had surmised—linking the Baraboo mountains with a continental-scale tectonic event that stretched from Labrador to Arizona.[9]

Still, the Baraboo Range, which emerges from the sea of sandstone like a leviathan just breaking the surface, holds many secrets. Geologic mystery is an infinite resource; there is no danger that it will ever be exhausted, and it's freely available to all who inquire. I ask the students to look again at the entire outcrop, now that they know the rocks' story better, making note of things they didn't see at first. Each time I come here, the rocks reveal something new; the eternal return brings deepening understanding. Today, the rocks are reminding me to be grateful for what has been lost and what has been found.

We're running behind schedule, but I decide to add a stop to our itinerary—a site just over the crest of the old Baraboo Range that shows how humanity may be able to heal itself after all. The place is the former site of a US Army facility, the Badger Army Ammunition Plant, where propellants for cannons, rockets, and small arms were manufactured from 1942 to 1975 for World War II and the wars in Korea and Vietnam. The "plant" actually consisted of hundreds of small cinder-block buildings scattered across an expanse of more than seven thousand acres and connected by a latticework of roads, a layout intended to prevent a fire or explosion in one area from compromising the entire installation. The result was a sprawling, dystopian "village" that, after the army closed it down, became a decaying ghost town with scores of poorly documented waste pits leaking PCBs and volatile organic compounds into the groundwater.

Beginning in the late 1990s, a consortium of hydrogeologists, botanists, environmentalists, local citizens, and members of the Ho-Chunk Nation—the indigenous people of the Baraboo area—began negotiations with the US Army to have the Badger Ammunition site decommissioned and remediated. Today, almost all of the buildings have been dismantled, and the grid of paved roads is slowly breaking up and becoming vegetated. About 75 percent of the site is now a state recreational area with bike trails, a restored prairie, and a vibrant community of supporters and volunteers. In 2015, the northwestern quarter, about 1,500 acres that lie snug against the southern edge of the Baraboo Hills, was signed over to the Ho-Chunk, marking the first time the US Army has ever returned land to a Native American tribe.[10] It's a tiny, largely symbolic step, but at least a start.

The sun is higher in the sky now; the earthy geosmin smell has dissipated, and with it the ghosts of long-ago springtimes. I'm back in the present and catch a whiff of onion—wild leeks, or ramps. I can't resist texting my husband to say the yearly foraging cycle is about to start. In my fifty-first winter, long after I had settled into a solitary life, an improbable thing happened: On a snowy trail, I met (well, skied past) a kindred Earthling with Norwegian roots. Our stratigraphies matched. He grew up on Cambrian sandstones along the Mississippi River in northeast Iowa. Like me, P had experienced divorce, the death of a spouse, and single parenthood, and had been worn down but managed to endure. We both have great unconformities in our lives, old eroded substrates covered incompletely by more recent events.

As a geotechnical contractor, P has a pragmatic relationship with rocks and happens to know the owners of every quarry and gravel pit in the state, a strategic alliance for me. He does carpentry, masonry, and plumbing—a good man for the apocalypse. Until then, we are content to let our lives revolve around the cycle of the seasons, starting with spring ramps, morels, and rhubarb, then a succession of wild berries (mul-, black-, rasp-, and blue-, in succession), the late summer bounty of garden vegetables, fall puffballs, hickory and hazelnuts, and winter snow for skiing.

I call out to the students that we will need to leave soon, and I smile to think how the group has been changed after only an hour together at this outcrop. These old rocks have cemented a loose association of classmates into a collective with durable bonds. Before we fold ourselves into the vans, I tell them the human story of Point of Rocks.

For many years, this outcrop was a hazardous place to stop, just off the side of US Highway 12. Coming from the south, the highway was ramrod straight where it passed the Badger Ammunition Plant, curved gently as it climbed up the flank of the Baraboo Hills, and then, just past the summit, veered sharply around the Point of Rocks promontory. In the early

2000s, as part of a larger highway improvement project, the US Department of Transportation announced plans to take the kink out of the road and blast the outcrop away. My students gasp, and I nod grimly in agreement. Geologists from all over the country were outraged; this was a place of pilgrimage. After one and a half billion years, was this to be the undignified demise of those sublime ripple marks? Led by Bob Dott, the connoisseur of quartz and one of my own mentors at the University of Wisconsin, then in his eighties, scores of us lobbied to have the outcrop spared. Ultimately, an overpass was built, removing the dangerous bend in the highway, and preserving Point of Rocks on a quiet cul-de-sac below.

Among the sadnesses of growing older is losing one's mentors, then, inexorably, one's peers—people who share memories of earlier versions of the world. But the Baraboo Quartzite reminds me that they're with us, in all the old familiar places. They're present when we immerse ourselves in secular ritual, on a spring visit to an outcrop, in the daily routine of classes, in the arc of another academic year, the seasons of a career, the slow overturn of the generations. I think of that lovely passage from George Eliot:

> We could never have loved the earth so well if we had had no childhood in it, if it were not the earth where the same flowers come up again every spring that we used to gather with our tiny fingers as we sat lisping to ourselves on the grass. . . . What novelty is worth that sweet monotony where everything is known and loved because it is known?[2]

And here comes old Luther again, still staking a claim on my thoughts: *What does Quartzite mean for us?* This time I may be closer to an answer. Quartzite means that while there will always be change, there will also be continuity; that the secular is sacred, and the ephemeral is eternal, so all we can do is get up every day and join the biogeochemical festivities, the sweet monotony, the ebb and flow, the sacraments of Earth and Life. Quartzite says, "Come forth from the depths, and once again behold the stars."[3]

EPILOGUE

Beach Stones

September again, south shore of Lake Superior in the Upper Peninsula of Michigan, not far from the Ontonagon River, where a tongue of glacial ice once dropped off a two-ton chunk of copper. The world's a mess. Pandemic, insurrection, polemic, and the hottest summer on record have left me feeling depleted and disoriented. I've come back to the Big Lake once more to find my center of gravity. As I walk toward the water's edge, my overheated thoughts are calmed by fingers of cool air and the white noise of waves washing over beach cobbles. This sound needs no processing through the filter of skepticism; there is no duplicity or hidden agenda. My mind relaxes, and I can hear the murmuring of the stones.

"Hello" I say, possibly aloud, happy that no other humans are within earshot. These cobbles feel like dear friends, even if my love for them is not reciprocated. There's such a variety here; in our once-glaciated part of the world, we hardly think twice about the fact that some rocks are travelers from distant places. I can guess where most of them came from and know their idiosyncratic stories, though each has new details to share. I feel lucky to have spent enough time in the company of rocks to understand their language. At this point, it's hard to remember when I didn't have rocks in my head. But my relationship with them has evolved and matured over time. We know so much more about this rocky planet than we did when I was a college sophomore picking up cobbles on the Lake Superior shore.

We now have a far more nuanced view of Earth's full biography, from simple basaltic beginnings through eons of self-reinvention. We can envision the tectonic dance of the continents and growth of mountain belts over time. We can simulate earthquakes and metamorphism in mathematical models. We can reconstruct the ghastly details of ancient supervolcano eruptions and asteroid impacts. We have begun to grasp the elaborate biogeochemical habits—the ceaseless interactions of rocks, life, water, and air—that maintain the world. And we have realized that through ignorance and insouciance, humans have altered the way the planet works, to our own peril.

At some point, rocks took control of my life. Fortunately, they have been wise counselors. Rocks have introduced me to the people I know best and granted us moments of *hierophany*, the sacred breaking through into everyday life. Rocks have taken my children and me around the world together and helped us see the whole planet as home. Rocks have taught me to value endurance over novelty; to perceive animacy in stillness and messages in silence; to respect the power of the incremental and accept the potential for the catastrophic; to be comfortable with unresolved mystery and unrequited affection.

My stone-strewn path through life may be anomalous, but even people who never give a thought to rocks are entirely under their jurisdiction. We are creatures shaped by the planet's rocky logic. Each of us is, most fundamentally, an *Earthling*.

On the beach, pebbles of Ordovician dolomite prattle with

Archean granite, their combined memories spanning half the age of the Earth. I wonder: What if people from different generations could gather this way? What would we learn if we could fold time back on itself and meet ancestors as contemporaries—or revisit moments in our own past? I yearn to embrace my little boys once more; to see my parents young and strong again; to remember how the world felt when I was a child. Rocks assure us that the past is no less real than the present. I spot a walnut-size piece of porphyritic basalt—one of the "Chinese calligraphy" stones my sister and I collected in childhood. I thank it for revealing itself to me and slip it into my pocket.

The stones are communicating with one another, with the waves and wind, with my feverish brain. A recent theory of consciousness[1] posits that intelligent awareness can emerge when the components of a large system have a certain level of interconnectivity. Neurons in the human brain reach the critical threshold. In the presence of these chattering cobbles, it seems obvious to me that, according to that definition, Earth is hyperconscious.

The connections among the many parts of Earth are like an elaborate neural network. Subduction of basalt keeps the interior and exterior of the planet in constant communication. Volcanoes propel subterranean magmas into the stratosphere as ash. Microbes repurpose volcanic carbon dioxide as shells. Sediments that tumbled into the ocean's abyss can be raised into alpine summits. Rocks deep underground swallow rainwater, a potion that catalyzes metamorphism. An ancient mountain belt buried by sand for half a billion years may find its way back to the surface. Glaciers gather rocks from across continents; on this beach, stones from the vast Canadian

Shield were brought together by ice for a symposium that is still underway twelve thousand years later.

All my old mentors are here. Sandstone, exquisite distillation, aquifer extraordinaire, favorite of white pines and stone masons. Basalt, common and unpretentious but willing to give its life for the tectonic good of the planet. There's drab dolomite, quietly storing carbon dioxide. Granite, representing the continental crust, and turbidites, who recycle it. Tuff and tektites to remind us that the world can change in a day, and flint and quartzite who teach us what it means to endure.

I close my eyes to listen and feel a wave of gratitude for the irrepressible creativity of this planet—all the surf crashing, rifts opening, carbon cycling, glaciers sliding, plates subducting, mountains rising, sediment accumulating, and rocks reincarnating. Earth doing what Earth does. What a place to grow up.

It's getting dark; time to go home. My pockets are full of pebbles, and there is a lot of sand in my socks.

ACKNOWLEDGMENTS

Rocks loom large in this book and partly obscure the many people who helped it take shape. The raw materials were provided by scores of mentors, colleagues, and students with whom I have shared a lifetime of memorable adventures in rock-strewn places around the world. The book began to crystallize with the encouragement of my agent, Eric Henney, and solidified under the wise guidance of my editor, Lee Oglesby. Copy editor Shelly Perron polished everything up, and Haley Hagerman produced illustrations that bring the stones to vivid life. All the while, my family has infused this rocky journey with joy.

NOTES

1. Sandstone

1. Buckley, E. R. *On the Building and Ornamental Stones of Wisconsin* (Madison: Wisconsin Geological and Natural History Survey, 1898), 230.
2. Irving, R. D. "On the classification of the Cambrian and Precambrian Formations," in *Seventh Annual Report of the United States Geological Survey*, J. W. Powell, ed. (Washington, DC: US Government Printing Office, 1888), 371–448.
3. Steen-Adams, M., Langston, N., and Mladenoff, D. "Logging the Great Lakes Indian reservations: The case of the Bad River Band of Ojibwe." *American Indian Culture and Research Journal* 34 (2010), 41–66.
4. Fitzpatrick, F., and Knox, J. "Spatial and temporal sensitivity of hydrogeomorphic response and recovery to deforestation, agriculture and floods in Wisconsin." *Physical Geography* 21, no. 2 (2000), 89–108.
5. Sloss, L. L. "Sequences in the cratonic interior of North America." *Geological Society of America Bulletin* 74 (1963), 93–114.
6. Dott, R. H. "The importance of eolian abrasion in supermature quartz sandstones and the paradox of weathering on vegetation-free landscapes." *Journal of Geology* 111 (2003), 387–405.

2. Basalt

1. Birks, J. B. *Rutherford at Manchester* (John Benjamins Publishing Company: New York, 1962), 108. Variations on this

quote have been used by other physicists, most notably Luis Alvarez in 1980 when paleontologists raised objections to his then-unverified hypothesis that the dinosaur extinction was caused by an asteroid impact. The remark only strengthened resistance to the idea in the paleontological community.

2. Chase, C., and Gilmer, T. "Precambrian plate Tectonics: The Midcontinent Gravity High." *Earth and Planetary Science Letters* 21 (1973), 70–78. https://doi.org/: 10.1016/0012–821X(73)90226–4.

3. Behrendt., J., et al. "GLIMPCE Seismic reflection evidence of deep-crustal and upper-mantle intrusions and magmatic underplating associated with the Midcontinent Rift system of North America." *Tectonophysics* 173 (1986), 595–615. https:/doi.org/doi: 10.1016/0040–1951(90)90248–7.

4. Pompeani, D., et al. "On the timing of the old Copper Complex in North America: A comparison of radiocarbon dates from different archeological contexts." *Radiocarbon* 63 (2021), 513–31. https://doi.org/10.1017/RDC.2021.7.

5. Redix, E. "'Our Hope and Our Protection': *Miskobiiwaabik* (Copper) and Tribal Sovereignty in Michigan." *American Indian Quarterly* 41 (2017), 224–49. https://doi.org/10.5250/amerindiquar.41.3.0224.

6. Schoolcraft, H. (*American Journal of Science* 3, 1821), 201–16. An account of the Native Copper on the southern shore of Lake Superior, with historical citations and miscellaneous remarks, in a report to the Department of War.

7. Krause, D. *The Making of a Mining District: Keweenaw Native Copper 1500–1870* (Detroit: Wayne State University Press, 1992), 15.

8. Palmer, W. *Strike in the Copper Mining District of Michigan* (Washington, DC: US Department of Labor, 1914), Library of Congress Control Number 14030170. https://lccn.loc.gov/14030170.

9. Lankton, L. *Hollowed Ground: Copper Mining and Community Building on Lake Superior, 1840s–1990s* (Detroit: Wayne State University Press, 2010), 216–225.

10. See, for example, Kellend, M., et al. "Increased yield and CO_2 sequestration potential with the C4 cereal sorghum bicolor cultivated in basaltic rock dust-amended agricultural soil." *Global Change Biology* 26 (2020), 3658–3676. https://doi.org/10.1111/gcb.15089.

3. Tuff

1. Hildreth, W., and Wilson, C. "Compositional zoning in the Bishop Tuff." *Journal of Petrology* 48 (2007), 951–99. https://doi.org/10.1093/petrology/egm007. *Also:* Hildreth, W., and Wilson, C. "The Bishop Tuff: New Insights from Eruptive Stratigraphy." *Journal of Geology* 105 (1997), 407–39.

2. White, D. A., Roeder, D., Nelson, T., and Crowell, J. "Subduction." *Geological Society of America Bulletin* 81 (1970), 3431–432. https://doi.org/10.1130/0016–7606(1970)81[3431:S]2.0CO;2.

3. Bakun, W., and McEvilly, T. "Earthquakes near Parkfield, California: Comparing the 1934 and 1966 sequences." *Science* 205 (1979), 1375–377. https://doi.org/10.1126/science.205.4413.1375.

4. Bakun, W., and Lindh, A. "The Parkfield, California earthquake prediction experiment." *Science* 229 (1985), 619–24. https//doi.org/10.1126/science.229.4714.619.

5. Ammon, C., et al. "Rupture process of the Sumatra-Andaman earthquake." *Science* 308 (2005), 1133–139. https//doi.org/10.1126/science.1112260.

6. Fei, H., and Katsura, T. "High water solubility of ring-woodite at mantle transition zone temperature." *Earth and Planetary Science Letters* 532 (2020). http//doi.org/10.1016/j.epsl.2019.115987.

7. Cooper, K., and Kent, A. "Rapid remobilization of magmatic crystals kept in cold storage." *Nature* 506 (2014), 480–83. http//doi.org/10.1038/nature12991. *Also* Woodward, A. "Pinpointing the trigger behind Yellowstone's last supereruption." *Eos* 98 (2017). http//doi.org/10.1029/2017EO065953.

4. Diamictite

1. Chamberlin, T. C. "The method of multiple working hypotheses." *Journal of Geology* 5 (1897), 837–48.

2. Murdoch, I. *The Sovereignty of Good* (London: Routledge & Kegan Paul, 1970), 52.

3. See, for example, Hoffman, P., et al. "Snowball Earth climate dynamics and Cryogenian geology-geobiology." *Science Advances* 3 (2017). http//doi.org/10.1126/sciadv.1600983.

4. Bjornerud, M. "Stratigraphic record of Neoproterozoic ice sheet collapse: The Kapp Lyell diamictite sequence, SW Spitsbergen, Svalbard." *Geological Magazine* 147 (2010), 380–90. http//doi.org/10.1017/S0016756809990690

5. Turbidite

1. Updike, J. "Satan's Works and Silted Cisterns" (review of Abdelrahman Munif's *Cities of Salt*). *The New Yorker*, October 17, 1988.

2. Anderson, M., and Kingsley, M. "Distribution and abundance of muskoxen and Peary caribou on Graham, Buckingham, and southern Ellesmere Islands in March 2015." *Rangifer* 37 (2017), 97–114. http//doi.org/10.7557/2.37.1 .4269.

3. Amy, R., Bhatnagar, R., Damkjar, E., and Beattie O. "The last Franklin expedition: report of a postmortem examination of a crew member." *Canadian Medical Association Journal* 135 (1986), 115–17.

4. Dott, R. H., Jr. "The geosynclinal concept." *SEPM Special Publication* 19 (1974), 1–12. http//doi.org/10.2110/pec.74 .19.

5. Walker, R. "Mopping up the turbidite mess," *Evolving Concepts in Sedimentology*, Robert Ginsburg, ed. (Baltimore: Johns Hopkins University Press, 1973), 28.

6. Karig, D., and Sharman, G. "Subduction and accretion in trenches." *Geological Society of America Bulletin* 86 (1974), 377–89.

7. Fryer, P., Ambos, E., and Hussong, D. "Origin and emplacement of Mariana forearc seamounts." *Geology* 13 (1985), 774–77. http//doi.org/10.1130/0091–7613(1985)1 3<774:OAEOMF>2.0.CO;2.

8. Scholl, D., and von Huene, R. "Crustal recycling at modern subduction zones applied to the past—Issues of growth and preservation of continental basement crust, mantle geochemistry, and supercontinent reconstruction." *Geological Society of America Memoirs* 200 (2007), 9–32.

9. Hawkesworth, C., Cawood, P., and Dhuime, B. "Rates of generation and growth of the continental crust." *Geoscience Frontiers* 10 (2019), 165–73. http//doi.org/10.1016/j.gsf .2018.02.004.

6. Dolomite

1. Ovid. *Metamorphoses* (8 CE), xv.

2. Dolomieu, D. *"Sur Un Genre de Pierres Calcaires Très-Peu Effervescentes Avec Les Acides"* ("On a Type of Limestone That Is Not Very Effervescent with Acids"). *Journal Physique* 39 (1791), 3–10.

3. Zenger, D., Bourrouilh-Le Jan, F., and Carozzi, A. "Dolomieu and the discovery of dolomite," in *Dolomite: A Volume in Honour of Dolomieu*, B. Purser, M. Tucker, and D. Zenger, eds. (London: Blackwell, 1995), 21–28.

4. McKenzie, J. "The dolomite problem: An outstanding controversy," in *Controversies in Modern Geology*, D. Moiler, J. McKenzie, and H. Weissert, eds. (London: Academic Press, 1991), 37–54.

5. For example, Burns, S., and Baker, P. "A geochemical study of dolomite in the Monterrey Formation, California." *Journal of Sedimentary Petrology* 57 (1987), 128–39.

6. Gregg, J., Bish, D., Kaczmarek, S., and Machel, H. "Mineralogy, nucleation and growth of dolomite in the laboratory and sedimentary environment: A review." *Sedimentology* 62 (2015), 1749–769. http//doi.org/10.1111/sed.12202.

7. Daye, M., Higgins, J., and Bosak, T. "Formation of ordered dolomite in anaerobic photosynthetic biofilms." *Geology* 47 (2019), 509–12. http//doi.org/10.1130/G45821.1.

8. Hazen, R., and Morrison, S. "On the paragenetic modes of minerals: A mineral evolution perspective." *American Mineralogist* 107 (2022), 1262–287. http//doi.org/10.2138/am-2022–8099.

7. Granite

1. Young, D. *Mind over Magma* (Princeton, NJ: Princeton University Press, 2002), 52.

2. Playfair, J. "Biographical account of James Hutton, M.D.,"
   in *The Works of John Playfair, Esq.*, vol. 4. (Edinburgh: Archibald Constable, 1822), 74–75.

3. Fox Keller, E. *A Feeling for the Organism: The Life and Work of Barbara McClintock* (New York: W. H. Freeman, 1983).

4. Young, D. *Mind over Magma*, 352.

5. Holmes is less well-known, even to geologists, than he should be. Biographer Cherry Lewis attempted to correct that with her book *The Dating Game* (Cambridge, UK: Cambridge University Press, 2002).

6. Read, H. H. "Granites and granites," in *Origin of Granite*, *Geological Society of America Memoir* 28 (1948), 1–20.

7. Reynolds, D. "The granite controversy." *Geological Magazine* 84 (1947), 209–23.

8. Lindsley, D. "Hard rock history." *Science* 302 (2003), 1334. http//doi.org/10.1126/science.1091919.

9. See, for example, Sizova, E., Gerya, T., Stüwe, K., and Brown, M. "Generation of felsic crust in the Archean, a geodynamic modeling perspective." *Precambrian Research* 271 (2015), 198–224. http//doi.org/10.1016/j.precamres.2015.10.005.

10. Johnson, T., Kirkland, C., Lu, Y., Smithies, R. H., Brown, M., and Hartnady, M. "Giant impacts and the origin and evolution of continents." *Nature* 608 (2022), 330–35. http//doi.org/0.1038/s41586-022-04956-y.

8. Eclogite

1. Mason, B. *Victor Moritz Goldschmidt: Father of Modern Geochemistry* (Washington, DC: The Geochemical Society, 1992).

2. Geochemical Society. "Goldschmidt: Father of modern geochemistry." *Elements* 13 (2017), 280–81. www .geochemsoc.org/honors/awards/vmgoldschmidtaward /victor-moritz-goldschmidt.

3. Einstein, A. *"Eine neue bestimmung der moleküldimensionen"* ("A New Determination of Molecular Dimensions"). *Annalen der Physik* 19 (1906), 289–306.

4. Jackson, S. W. "Why are the ashes of one of Norway's most important scientists in an unmarked grave?" *Science Norway* (2020). https://sciencenorway.no/science-history/why-are -the-ashes-of-one-of-norways-most-important-scientists -in-an-unmarked-grave/1764162.

5. Centola, D., Becker, J., Brackbill, D., and Baronchelli, A. "Experimental evidence for tipping points in social convention." *Science* 360 (2018), 1116–119. http//doi.org/10 .1126/science.aas8827.

9. Glass and Flint

1. Kirkpatrick, J., Fagereng, Å., and Shelley, D. "Geologic constraints on the mechanisms of slow earthquakes." *Nature Reviews* 2 (2019), 285–301. http//doi.org/10.1038 /s43017-021-00148-w.

2. Longino, H. "Cognitive and non-cognitive values in sciences: Rethinking the dichotomy," in *Feminism, Science, and the Philosophy of Science*, L. Nelson and J. Nelson, eds. (Amsterdam: Kluwer Publishers, 1996), 39–58.

3. Abbott, A., and Nosengo, N. "Italian seismologists cleared of manslaughter." *Nature* 515 (2014), 171. http//doi.org/10 .1038/515171a.

4. Mitchell, R., et al. "Oceanic anoxic cycles? Orbital prelude to the Bonarelli Level (OAE2)." *Earth and Planetary Sci-*

ence Letters 267 (2008), 1–16. http//doi.org/10.1016/j.epsl
.2007.11.026.

5. Wierer, U., et al. "The Iceman's lithic toolkit: Raw material,
technology, typology and use." *PLOS One* 13, no. 6 (2018):
e0198292. http//doi.org/ 10.1371/journal.pone.0198292.

6. Alvarez, L., Alvarez, W., Asaro, F., and Michel, H. "Extraterrestrial cause for the Cretaceous-Tertiary extinction."
Science 208 (1980), 1095–108. http//doi.org/10.1126
/science.208.4448.10.

7. Smit, J. "The global stratigraphy of the Cretaceous-Tertiary
boundary impact ejecta." *Annual Reviews of Earth and
Planetary Sciences* 27 (1999), 75–113.

8. Brown, M. "The Debate over Dinosaur Extinctions Takes
an Unusually Rancorous Turn," *New York Times*, January
19, 1988: C-1.

9. See, for example, Keller, G., et al. "Mercury linked to Deccan
Traps volcanism, climate change and the end-Cretaceous
mass extinction." *Global and Planetary Change* 194 (2020),
1–17. http//doi.org/10.1016/j.gloplacha.2020.103312.

10. Bosker, B. "The Nastiest Feud in Science." *The Atlantic*
(September 2018).

11. Barghoorn, E., and Tyler, S. "Microfossils from the Gunflint Chert." *Science* 147 (1965), 563–75. http//doi.org/10
.1126/science.147.3658.

12. Addison, W., et al. "Discovery of distal ejecta from the
1850 Ma Sudbury impact." *Geology* 33 (2005), 193–96.
http//doi.org/10.1130/G21048.1.

13. Jirsa, M., Fralick, P., Weiblen, P., and Anderson, J. "Sudbury impact layer in the western Lake Superior region."
Geological Society of America Field Guide 24 (2011), 147–69.
http//doi.org/10.1130/2011.0024(08).

14. Addison, W., et al. "Debrisites from the Sudbury impact event in Ontario, north of Lake Superior, and a new age constraint: Are they base-surge deposits or tsunami deposits?" in *Large Meteorite Impacts and Planetary Evolution IV*, R. Gibson and W. Reimold, eds. *Geological Society of America Special Paper* 456 (2010). http//doi.org/10.1130/2010.2465(16).

15. State of Wisconsin, 2013. Act I. docs.legis.wisconsin.gov/2013/related/acts/1.pdf. In the end, the proposed mine was not developed; the iron ore was too low grade to make mining it economical. The entire proposal may have been a political strategy to relax regulations for future mining projects in the state.

10. Quartzite

1. Eliade, M. *The Sacred and the Profane*. W. R. Trask, trans. (New York: Harcourt, Brace & World, 1957), 68–69, 94.

2. As Stewart Brand famously wrote in the preface to the *Whole Earth Catalog* in 1968, "*We* are as *gods* and might as well *get* good at it."

3. Vance, M. *Charles Richard Van Hise: Scientist Progressive*. (Madison: Wisconsin Historical Society Press, 1960).

4. Vance, 83–87.

5. Paulhus, D., and Williams, K. "The dark triad of personality: Narcissism, Machiavellianism, and psychopathy." *Journal of Research in Personality* 36 (2002), 556–63. http//doi/10.1016/S0092–6566(02)00505–6.

6. As articulated by Milton Friedman in his influential essay "The Social Responsibility of Business Is to increase Its Profits." *New York Times*, September 13, 1970: SM-17.

www.nytimes.com/1970/09/13/archives/a-friedman -doctrine-the-social-responsibility-of-business-is-to.html.

7. Book of Numbers 20: 2–13.
8. Stewart, E., Brengman, L., and Stewart, W. "Revised Provenance, Depositional Environment, and Maximum Depositional Age for the Baraboo (<ca. 1714 Ma) and Dake (<ca. 1630 Ma) Quartzites, Baraboo Hills, Wisconsin." *Journal of Geology* 129 (2021). http//doi.org/10.1086 /713687.
9. Medaris, L. G., et al. "Early Mesoproterozoic evolution of midcontinental Laurentia: Defining the Geon 14 Baraboo orogeny." *Geoscience Frontiers* 12 (2021). http//doi.org/10 .1016/j.gsf.2021.101174.
10. Sauk Prairie Conservation Alliance. https://saukprairievision .org/.

Epilogue

1. Koch, C. "What is consciousness?" *Nature* 559 (2018), S8-S12. https://www.nature.com/articles/d41586-018 -05097-x
2. Eliot, George. *The Mill on the Floss* (New York: Bantam Books, 1987).
3. To paraphrase the last line of Dante's *Inferno*.

INDEX

Abruzzo, Italy, 240
Acasta Gneisses, 197
accreted terrane, 126
accretionary prisms, ix, 139–44, 147
active-lid convection, 67, 207
Addison, Bill, 252–55, 256
agates, 16, 49, 59
agriculture, 14, 17, 20, 23–24, 28–29,
    38–39
Aleutian Islands, 79
algae, 17, 36, 158
Alpine Fault, 229–30, 232
Alps, 135–36, 158–59, 184, 244
aluminum, 64
Alvarez, Luis, 248–49, 256
Alvarez, Walter, 247–50, 256
Amatrice earthquake, 225
*American Association of Petroleum Geologists
    Bulletin*, 33
amygdaloidal basalts, 49, 59
Andes, 79, 124, 194
Andesite magmas, 81
Annual Report of the States Geological Survey
    (Powell), 19–20
anoxic events, 245–46
Answers in Genesis, 170
Antarctica, 110, 115, 121, 125, 204
Anthropocene, 14, 23, 245
antisemitism, 210, 220–21
Apennines, 228, 240, 242–49
Appalachian-Caledonian belt, 1–2, 99,
    207–8
Appalachians, 116, 135, 169
aquifers, 9, 30–31, 34, 38, 282
Archaeological Museum (Bolzano), 256
archaeology, 59–60, 129
Archean time, 196, 281
Arctic, 2, 4–5, 94–111, 118–19, 122, 128
Arendt, Hannah, 175
Armstrong, Neil, 66
asbestos, 55
astronomical cycles, 242, 246
Atlantic Ocean, 68
atmosphere, 32, 46, 49, 89, 93, 158, 207

atomic bomb, 249
atoms, 5
    diffusion of, 212
augite, 189
Australia, 197, 234, 254
Australian plate, 229

bacteria, 31, 165
bacterial fossils, 254
Badger Army Ammunition Plant,
    273–75
Bad River Ojibwe reservation, 258
Baltic Shield, 184
Banda Aceh earthquake of 2004, 87
Baraboo Hills, 20, 260–76, *262*
basalt, viii, *viii*, 42–72, *43*, 70, 79–80, 92,
    146, 178, 188, 195, 198, 205–7, 215,
    280–82
basaltic shield volcanoes, 66
Basin and Range region, 116
batholiths, 183
bedding planes, 184, 269–70
Beechey Island, 129
Bellsund Fjord, 94, 99, 107, 119
Beltrami, Giacomo, 241, 253, 256
biogenic methane, 118
biogenic mineral species, 166
biogeochemical cycles, 163, 166, 266, 280
biogeochemistry, 165–66
biosphere, 266
biotite, 189
Birds, The (film), 110
Bishop Tuff, 77–78, 82, 91
black smokers, 80
Blade Runner (film), 71
Bonarelli level, 245–47
Boundary Waters, 40
Bowen, Norman, 187–91, 194, 201
Bowen's reaction series, 188–89, 196
brachiopods, 148
brittle-ductile transition, 212
Bronze Age, 244, 256
Brumpton, Greg, 252–55
bryozoa, 148

ABOUT THE AUTHOR

Marcia Bjornerud is a professor of environmental studies and geosciences at Lawrence University. She is a contributing writer to the *New Yorker, Wired, Wall Street Journal,* and *Los Angeles Times* and the author of *Reading the Rocks, Timefulness,* and *Geopedia.*